4th International Conference on Mechatronics and Electrical Systems (ICMES 2019)

IOP Conference Series: Materials Science and Engineering
Volume 841

Prague, Czech Republic
11 - 13 October 2019

ISBN: 978-1-7138-2683-5
ISSN: 1757-8981

Printed from e-media with permission by:

Curran Associates, Inc.
57 Morehouse Lane
Red Hook, NY 12571

Some format issues inherent in the e-media version may also appear in this print version.

Institute of Physics
Dirac House, Temple Back
Bristol BS1 6BE UK

Phone: 44 1 17 929 7481
Fax: 44 1 17 920 0979

techtracking@iop.org

Additional copies of this publication are available from:

Curran Associates, Inc.
57 Morehouse Lane
Red Hook, NY 12571 USA
Phone: 845-758-0400
Fax: 845-758-2633
Email: curran@proceedings.com
Web: www.proceedings.com

TABLE OF CONTENTS

MECHATRONICS AND ELECTRICAL ENGINEERING

MECHANICAL ENGINEERING AND APPLIED MECHANICS

IOP Conf. Series: Materials Science and Engineering **841** (2020) 011001 doi:10.1088/1757-899X/841/1/011001

Preface

It is our pleasure to welcome you to the 2019 4th International Conference on Mechatronics and Electrical Systems (ICMES 2019) in Prague, Czech Republic on October 11-13, 2019. ICMES 2019 is a conference dedicated to topics in advances in aerospace and automotive, applied sciences and biosciences, biomechanics and medical technology, dynamic, vibration, acoustic and control system, control, robotics and mechatronics, artificial intelligence and intelligent control, etc.

The proceedings present a selection of 13 papers submitted to the conference from universities, research institutes and industries. All of the papers were subjected to peer-review by conference committee members and international reviewers. Papers included in the proceedings have been selected according to quality and relevance to the conference themes. The proceedings aim to present to the readers recent advances in the field of mechatronics and control engineering and in various related areas.

We would like to express our genuine gratitude to everyone who has contributed to ICMES 2019. The opening remark including a warm welcome to all participants has been given by Assoc. Prof. Jan Faigl, Czech Technical University (CTU), Czech. Our especially thanks to the international advisory committee, Prof. Chun-Yi Su, Concordia University, Canada, the program chairs, Prof. Juan M. Corchado, University of Salamanca, Spain, Assoc. Prof. Ahmed Abdelgawad, Central Michigan University, USA, all the technical committee members. Let us wish ICMES the same success for next year.

Conference Committees

IOP Conf. Series: Materials Science and Engineering **841** (2020) 011002 doi:10.1088/1757-899X/841/1/011002

Conference Committees

International Advisory Committee

Chun-Yi Su, Concordia University, Canada

Conference Chair

Jan Faigl, Czech Technical University in Prague, Czech

Conference Program Chairs

Juan M. Corchado, University of Salamanca, Spain

Ahmed Abdelgawad, Central Michigan University, USA

Yiyong Yang, China University of Geosciences (Beijing), China

Technical Committees

Sanjeet Dwivedi, Danfoss Power Electronics, Denmark

Paulo Nobre Balbis dos Reis, University of Beira Interior, Portugal

Xinggang Yan, University of Kent, UK

Xiaonlin Wang, University of Tasmania, Australia

Mario Chauca, Ricardo Palma University, Peru

Bandit Suksawat, King Mongkut's University of Technology North Bangkok, Thailand

Jr-Rung Chen, Industrial Technology Research Institute, Taiwan

Sanjay Deokar, Zeal College of Engineering and Research, India

Kemal Ucak, Mugla Sitki Kocman University, Turkey

Evgeny Blagin, Samara National Research University, Russia

Mohd Azlan MohdIshak, Universiti Teknologi MARA, Malaysia

Ray Hwa Wong, Hwh Hsia University of Technology, Taiwan

Mohamed Arezki Mellal, M'Hamed Bougara University (UMBB), Algeria

Mohamed bin Hussein, Universiti Technologi Malaysia, Malaysia

IOP Conf. Series: Materials Science and Engineering **841** (2020) 011003 doi:10.1088/1757-899X/841/1/011003

First name	Surname	Institution
Tsung-Han	Hsieh	Industrial Technology Research Institute, Taiwan
H T	Thakker	Pandit Deendayal Petroleum University, India
B	Bedkowski	Institute of Electrical Drives and Machines KOMEL, Katowice, Poland
P	Baranov	National Research Tomsk Polytechnic University, Russia
Yu V	Belousov	Bauman Moscow State Technical University, Russian Federation
Bowen	Hou	Beijing Jiaotong University, China
Juan Carlos Valencia	Salinas	Universidad de las Fuerzas Armadas ESPE, Ecuador
K	Karpukhin	NAMI Russian State Scientific Research Center, Russia
Phuoc Vinh	Dang	The University of Danang – Danang University of Science and Technology, Vietnam
Chang-pu	MA	Tongji University, China; Guilin University of aerospace technology, China
Shigeru	Sakuta	Institute of Technologists, Japan
Pandeng	Zhang	Tongji University, China
Charlyn May A	Gapuz	Mapua University, Philippines

IOP Conf. Series: Materials Science and Engineering **841** (2020) 011003 doi:10.1088/1757-899X/841/1/011003

E-mail address

HenryHsieh@itri.org.tw

harshit041196@gmail.com

b.bedkowski@komel.katowice.pl

bpf@tpu.ru

belou.80@mail.ru

bwhou@bjtu.edu.cn

ajcvalencia5@espe.edu.ec

K.Karpukhin@nami.ru

dpvinh@dut.udn.vn

smcp1108@foxmail.com

m01411129@iot.ac.jp

zhangpandeng@tongji.edu.cn

cha.a.gapuz@gmail.com

IOP Conf. Series: Materials Science and Engineering **841** (2020) 011004 doi:10.1088/1757-899X/841/1/011004

Peer review statement

All papers published in this volume of *IOP Conference Series: Materials Science and Engineering* have been peer reviewed through processes administered by the proceedings Editors. Reviews were conducted by expert referees to the professional and scientific standards expected of a proceedings journal published by IOP Publishing.

IOP Conf. Series: Materials Science and Engineering **841** (2020) 012001 doi:10.1088/1757-899X/841/1/012001

Fundamental Study on Hysteresis Reduction of Piezoelectric Actuators by Preload Control

Shigeru Sakuta[1] **and Shigeomi Tsuruoka**[2]

[1] Department of Mechanical and Production Engineering, Institute of Technologists,333 Maeya Gyoda-City Saitama, 361-0038 JAPAN
[2] Kaneko Agricultural Machinery 1-516-10 Komatsudai Hanyu-city Saitama, 348-0038 JAPAN

Email: m01411129@iot.ac.jp

Abstract. Piezoelectric actuator is usually used in feedback system because of the hysteresis non-linearity of between input voltage and output displacement. The aim of this paper is analysis of the hysteresis to reduce cost of high precision positioning system by piezoelectric actuator with open-loop control. Ratio of max hysteresis to max displacement, RHD, has been proposed as an evaluation indicator for the hysteresis. Effects of compressive stresses, i.e. preload, in two directions on RHD were investigated and bulk and stacked piezoelectric actuators with low RHD have been studied experimentally.

1. Introduction

Piezoelectric actuator with less hysteresis can be applied to high precision open-loop control. Open-loop control is not costly, compared to closed-loop one. It is possible for less hysteresis piezoelectric actuator with open-loop control to reduce the cost of positioning systems [1]. In this paper, effects of a stress onto a bulk and stacked piezoelectric actuators on the hysteresis are studied. Unruan M et al. described preload to a piezo lessens hysteresis because of domain wall suppression [2] and Yimnirun R showed that dielectric properties were lowered by a stress [3]. New hysteresis indicator, RHD, was introduced to evaluate hysteresis of both a bulk and a stacked piezoelectric actuators. Preload effects on RHD have been studied for a bulk and stacked piezoelectric actuator. Although application of some control theories to piezoelectric actuators in order to reduce piezoelectric actuator's hysteresis nonlinearity have been studied [4][5][6], the control systems become complicated.

2. Experiment

2.1. Experimental Configuration

Figure 1 shows a configuration of an experiment equipment. Two kinds of specimens, i.e. a bulk and a stacked piezoelectric actuators, were used in the experiment. Size of each piezoelectric actuator is 10(width)×10(depth)×40mm(height). Table 1 shows the piezoelectric actuator specifications. The actuator was under the preload up to 4.9 [MPa] in X and Y direction, shown in figure 1, where X is a polarized direction of the piezoelectric actuator. A stacked type couldn't be at a large compressive stress in Y direction because of its configuration of stacked thin PZT plates in X direction. Actually the preload in Y-direction for a stacked piezoelectric actuator was limited up to 0.09 [MPa]. The preload was generated by two air cylinders in X and Y directions and the compressive stresses were applied to the actuator through floating joints. Max voltage to a bulk and a stacked piezo was 500[V] and 100[V] respectively. The voltage was applied through high voltage supply using ramp output function. Time of up and down ramp was 20 [s]. The input signal rate is small enough for piezoelectric actuators. Halt time between the ramps was 1[s]. Displacement of the bulk and the stacked piezoelectric actuator was measured by laser interferometer system (5519A, Agilent Technologies) with 5 nm resolution and electric micro (E-M5R, Tokyo Seimitsu) with sensitivity of 0.01 V/μm respectively. Figure 2 shows a block diagram of the experiment. Open-loop control was adopted with stresses in X and Y directions to evaluate hysteresis of piezoelectric actuator.

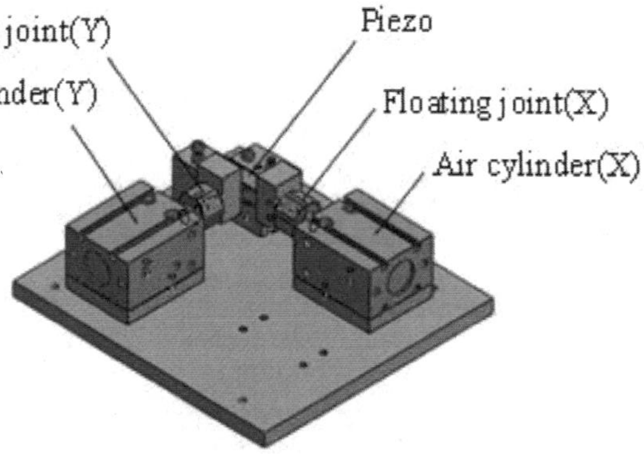

Figure1. Experimental configuration: Stress in X and Y directions were applied to a piezoelectric actuator using air cylinders

Table 1. Comparison of specimen

Parameter	Bulk	Stacked
Size (width x Depth x Height[mm])	10 x 10 x40	10 x 10 x40
Material	PZT	Stacked PZT
Sensitivity[V/μm]	3000	4
Response [kHz]	39	34
Manufacturer	Fuji Ceramics	NEC Tokin

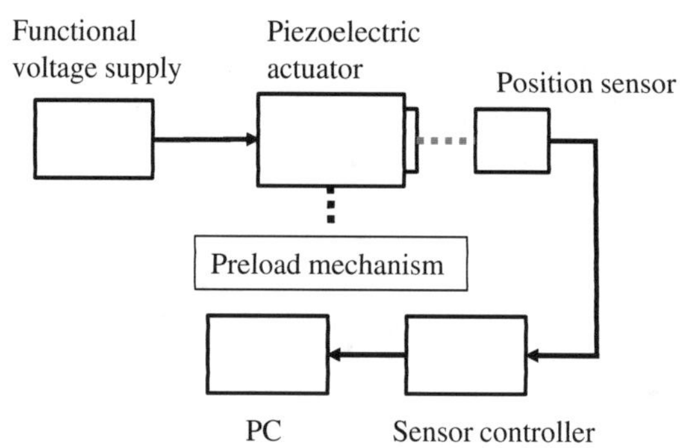

Figure 2. Block diagram of experiment apparatus: Open-loop control was adopted with stress in X and Y directions to evaluate hysteresis of piezoelectric actuator

2.2. RHD Definition

Figure 3 shows a relationship between input voltage and output displacement of a stacked piezoelectric actuator in X direction. Ratio of max hysteresis to max displacement, i.e. operating range, RHD, has been defined newly as an evaluation indicator for the hysteresis.

IOP Conf. Series: Materials Science and Engineering **841** (2020) 012001 doi:10.1088/1757-899X/841/1/012001

$$\mathrm{RHD} = \frac{\text{Max hysteresis}}{\text{Max displacement}} \tag{1}$$

It is difficult to compare hystereses of bulk and stacked piezoelectric actuators because there are large differences between operating ranges of bulk piezoelectric actuator and stacked one, for example, 30μm and 0.1μm respectively. RHD indicator makes it possible to compare hysteresis of bulk and stacked piezoelectric actuators. Since RHD is a kind of normalized estimate indicator, characteristics of hysteresis for bulk and stacked piezoelectric actuators can be discussed using RHD.

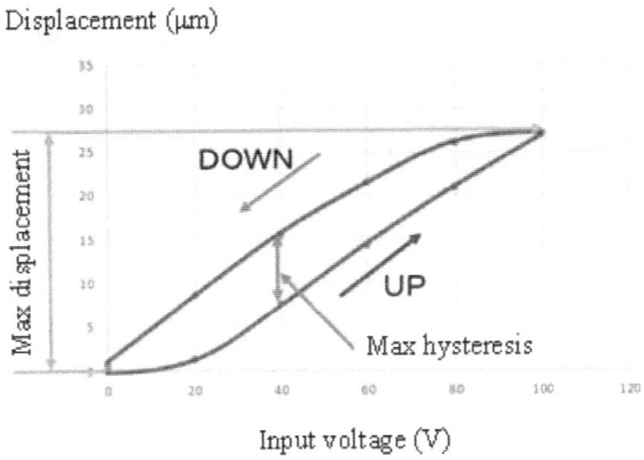

Figure 3. Definition of RHD: In order to compare hysteresis of both bulk and stacked piezoelectric actuator, RHD was defined as max-hyteresis/max-displacement

3. Experimental Results

Experimental data were analyzed statistically. Figure 4 shows results of RHD by multiple regression analysis. When the significance level was 0.05, the null hypothesis that RHD is not influenced by X and Y preload is rejected because the P values are smaller than 0.05. On the other hand, RHD was not affected by type of the piezoelectric actuators. Figure 5 and figure 6 shows relationship between preload and RHD in X and Y direction. Figure 5 shows that RHD decreases with an increase of X-preload. Y preload had effects on RHD too, however there were no clear trends like X-preload. . Type did not influence RHD statistically because statistics P value of type was more than 0.05.

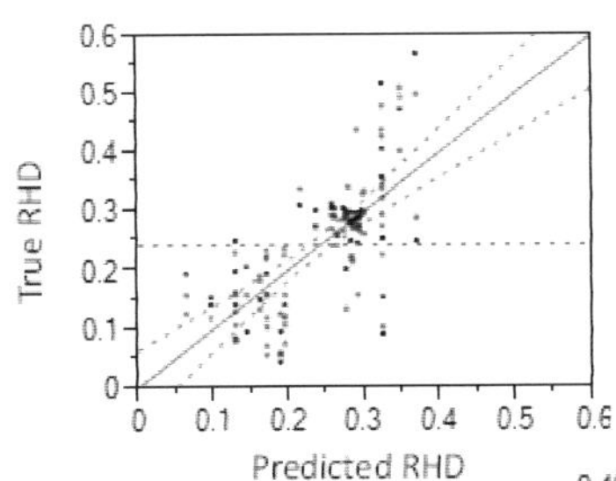

R-Sq	0.496649
R-Sq(adj)	0.484759

Predictor	Coef	StDev	T	P
Constant	0.3172934	0.015832	20.04	<.0001*
Type[Bulk]	0.0126802	0.012447	1.02	0.3103
X preload [MPa]	-0.091153	0.01069	-8.53	<.0001*
Y preload [MPa]	0.1106098	0.039518	2.80	0.0059*

Figure 4. Multiple regression analysis: Multiple regression analysis shows that X preload influenced RHD most

IOP Conf. Series: Materials Science and Engineering **841** (2020) 012001 doi:10.1088/1757-899X/841/1/012001

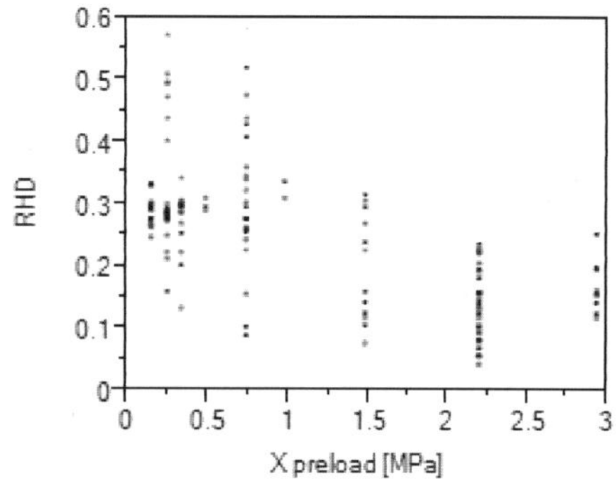

Figure 5. Relationship between X preload and RHD: There was a trend that RHD decreased as X preload increased

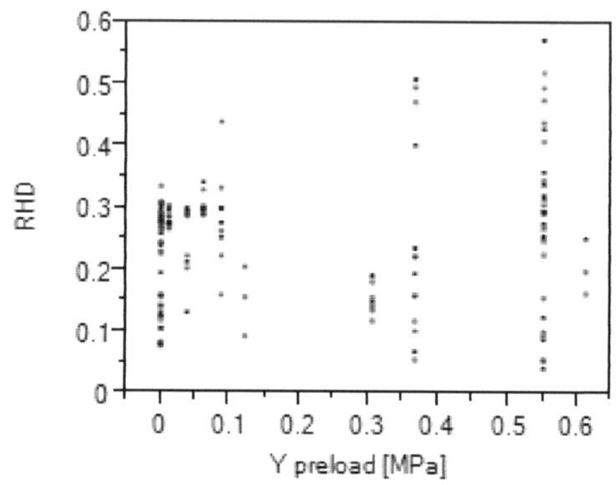

Figure 6. Relationship between Y preload and RHD: There was a not clear trend between Y preload and RHD

4. Discussion

Preload influences a max displacement as well as RHD. It is important to take into consideration actuator stroke from viewpoint of actual positioning specifications on positioning travel. Figure 7 shows a relationship between a max displacement and RHD. Displacement of stacked piezoelectric actuator was much larger than bulk's, so there are two groups in figure 7. RHD for bulk piezoelectric actuator was not clear in figure 7.

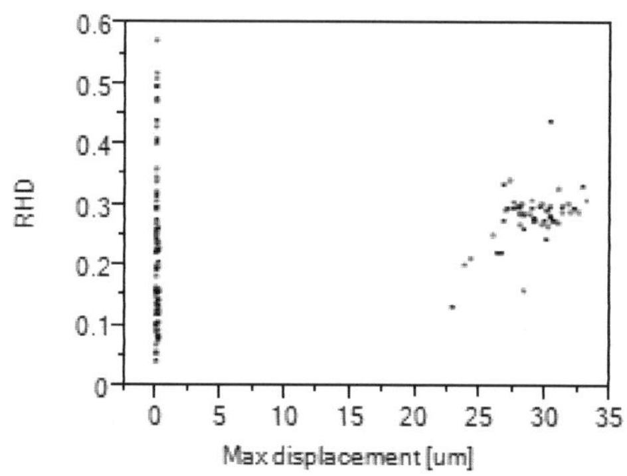

Figure 7. Relationships between max displacement and RHD: different displacement range between bulk and stocked piezoelectric actuator resulted in two groups

Preload influences a max displacement as well as RHD. It is important to take into consideration actuator stroke from viewpoint of actual positioning specifications on positioning travel. Figure 7 shows a relationship between a max displacement and RHD. Displacement of stacked piezoelectric actuator was much larger than bulk's, so there are two groups in figure 7. RHD for bulk piezoelectric actuator was not clear in figure 7.

To study RHD of bulk piezoelectric actuator in detail, the bulk data were only analyzed in figure 8. Both preload groups of High (more than 1.5 [MPa]) /Low X-preload, High (more than 1.2 [MPa]) /Low Y-preload, are plotted in figure 8. Points with high X-preload and high Y-preload data seem to cluster in A (figure 8) and points with high X-preload and low Y-preload seem to cluster in B. A has a condition of open-loop high accuracy positioning with low RHD and small stroke. B has a condition of open-loop control with a relatively large stroke.

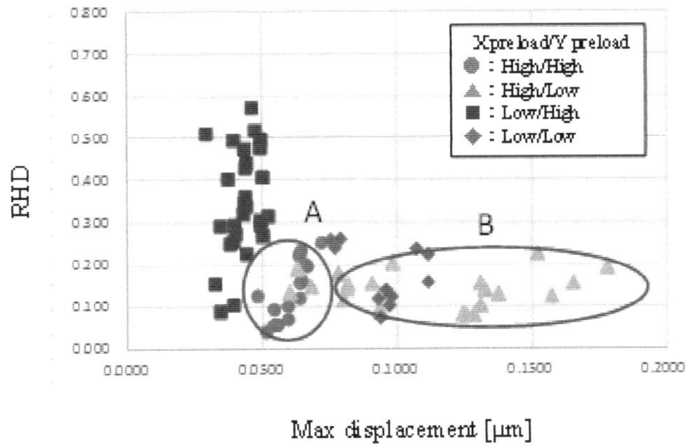

Figure 8. Relationships between max displacement and RHD for bulk piezoelectric actuator: Points with high X preload and high Y preload cluster in A and points with high X preload and low Y preload clustered in B

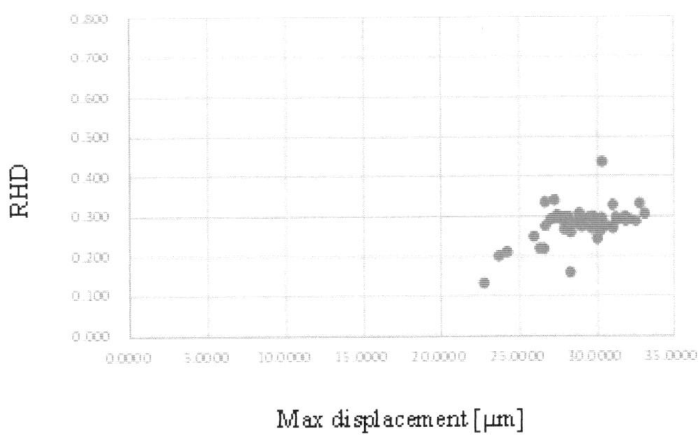

Figure 9. Relationships between max displacement and RHD for stocked piezoelectric actuator: Chart didn't agree with hyperbola chart, which shows that there was a complicate hysteresis is trend fo stacked piezoelectric actuator

Figure 9 shows relationship between max displacement and RHD of stacked data. It can be seen that all data cluster around one spot. There doesn't seem to be a clear trend in stacked piezoelectric actuators like bulk ones. If hysteresis is constant, figure 8 and figure 9 become hyperbola from the RHD definition (equation (1)). Figure 8 revealed that the relationship was approximately hyperbola, however, Figure 9 doesn't seem to be hyperbola. Figure 9 suggests that hysteresis in stacked piezoelectric actuator changed greatly, compared to those of bulk piezoelectric actuators. High X-preload tends to make RHD low for a stacked piezoelectric actuator, too. However, there was no clear relationship between displacement and RHD for a stacked piezoelectric actuator as shown in figure 9.

As stated in Chapter 2, large preload in Y-direction, for example more than 0.09 [MPa], can't be stressed to a stacked piezoelectric actuator. Limited preload in Y-direction onto a stacked piezo might cause the unclear. Nevertheless the stroke is very small, for example, less than 100 [nm], low RHD and small stroke condition, that is, large preloads in X and Y directions are good for high precision positioning with open-loop control because hysteresis error can be estimated from equation (1) and figure 8 as follows:

$$\text{Max hysteresis} \fallingdotseq \text{RHD} \times \text{Max displacement}=0.1\times 0.05\ [\mu\text{m}] = 5\ [\text{nm}] \qquad (2)$$

Positioning error due to hysteresis is evaluated approximately to be 5 [nm] with large preloads in X and Y directions.

Considering engineering applications, dynamics of a piezoelectric actuator with X and Y preload should be investigated in detail. For example, RHD of piezoelectric actuator in actual servo system should be looked into. Frequency response of a piezoelectric actuator with X and Y preload should be investigated, too. Especially the dynamics of a bulk piezoelectric actuator with large X preload, which has low RHD as shown in figure 8, are interesting.

5. Summary

New indicator for hysteresis, RHD, has been introduced to evaluate the hystereses of both a bulk and a stacked piezoelectric actuators. Experimental results demonstrated that high X and Y preload was good for precision positioning with open-loop control for a bulk piezoelectric actuator.

Although feasibility is influenced by required positioning accuracy, it can be seen that open-loop positioning by a bulk piezoelectric actuator can be applied to precision positioning with high preloads in X and Y directions. The authors are currently investigating dynamics of open-loop precision positioning on the condition.

6. References

[1] Sakuta S 2001 *Japanese Patent* 3256556
[2] Unruan M, Prasartketrakarn A, Ngamijarurojana A, Laosiritaworn Y and Ananta S, 2009 *J.Appl.Phys* 105
[3] Yimnirun R, Vuruan M, Laosiritaworn Y and Ananta S 2006 *Phys*.D **39** 3097
[4] Lin C.J and Yang S.R 2005 *Asian Journal of Control* **7** No.1 73
[5] Slen J.C., Jywe W.T, Chiang H.K and Shu Y.L 2007 *Proc. Of 15th Mediterranean on Control & Automation* T2-002
[6] Sung B.J, Lee E.W and Lee J.G 2008 *J. Electrical Engineering & Technology* **3** No.3 401

Acknowledgments

The authors would like to thank Mr.Yagihashi Y, Mr.Momoi N, Mr.Oka Y, Mr.Kamiya T, Mr.Suzuki K and Mr.Shimoda H for their assistance.

IOP Conf. Series: Materials Science and Engineering **841** (2020) 012002 doi:10.1088/1757-899X/841/1/012002

Portable Rotary-angle Detector for Industrial Motor Inspection

Tsung-Han Hsieh[1], Jr-Rung Chen[1], Po-Er Hsu[1] and Bing-Lin Ho[1]

[1] Center for Measurement Standards, Industrial Technology Research Institute, Taiwan

Email: JrRungChen@itri.org.tw

Abstract. Several types of precision equipment, such as machine tools and robot arms, rely on precision rotary motion. In this study, we developed a portable rotary-angle detector (PRAD) comprising five optical sensors and one angle ring with a 57-mm diameter to measure the positioning and radial errors of rotary motions. The five optical sensors were installed at equally divided intervals around the angle ring according to the equal division averaged method (EDA-Method) implementation. The grating of the angle ring has 9000 lines, and the signal interpolation of the angle detector system has 4000-fold subdivision. Therefore, the PRAD resolution can be achieved at 0.036 arcsec. Furthermore, a 12-face polygon and an autocollimator with 0.05-arcsec resolution and 0.25-arcsec accuracy were used to check the PRAD performance. The experimental results showed that the structure of the PRAD is simple. However, the difference over 360° is only 0.7 arcsec, indicating that PRAD is suitable for motor inspection in machine tools or robot arms.

1. Introduction

With the advance of industry development, robotic arms are widely used. Applications such as robotic machining, welding, and pick-and-place demand good accuracy of these robots. Another important application is in machine tools. This has been a challenge because the tool at the end must be positioned with extreme precision, especially in a five-axis machine tool. Rotary table have geometric errors in six degrees of freedom as described in ISO230-1 [1]. Some sensors are already preinstalled in the machine tool or mounted in the robot arm to capture the position of every joint. In the machine tool industry, this makes them ideal for monitoring spindle rotations per minute (RPM). The most commonly used sensors are angle encoders and rotary encoders [2]. Angle encoders detect the rotational relationship between two parts with high precision, typically with accuracy better than ±10″. Owing to such extreme levels of precision, angle encoders are important in the development of multi-axis centers. Regardless of simultaneous or independent movements, the designs of components such as the rotary table and swivel head often employ this type of encoder. Rotary encoders are similar to angle encoders, and measure rotation. However, they are not as precise as angle encoders. For angular calibration of a rotary encoder, this study aims to develop a convenient tool to quickly measure and evaluate angular positioning accuracy.

2. PRAD Development

For industrial application, a detector should be portable and convenient for engineers to use in assembly or production lines. Fast installation is a major consideration. Multi-read heads can be built as the comparative angle measurement equipment to calibrate the angle position error of two rotary encoders whose graduation position deviation are unknown by the equal division averaged (EDA) method [3–6]. Figure 1 shows the conceptual model of comparative angle-measurement equipment, which we call a portable rotary-angle detector (PRAD) system. As shown in Figure 2, the PRAD developed in this paper consists of five read heads to implement EDA. Angle α of the PRAD and angle β of the rotary angle of the actuator (to be measured or tested) are connected with the same shaft and rotate at constant speed. $\alpha(\theta)$ can be obtained by equation (1). $A_0(\theta)$ to $A_4(\theta)$ are the readout angles from the optical read heads. The number of residual errors was small because the eccentric error was eliminated by the EDA method. Therefore, the residual errors could be ignored in this study. The difference between the PRAD and motor can be obtained by equation (2) and is indicated as $\delta(\theta)$, which is also called the positioning error of the actuator in the PRAD system.

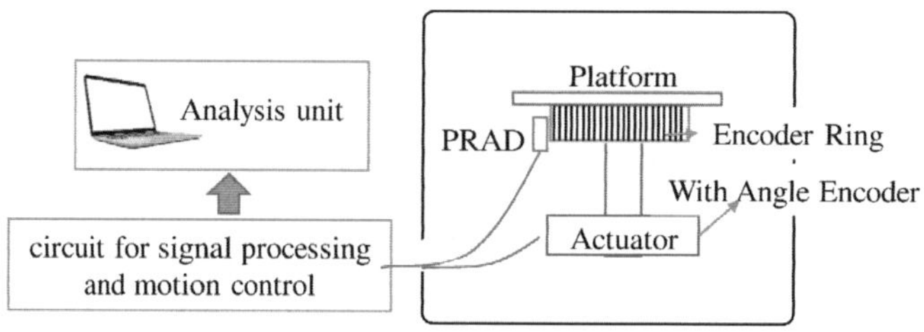

Figure 1. PRAD system

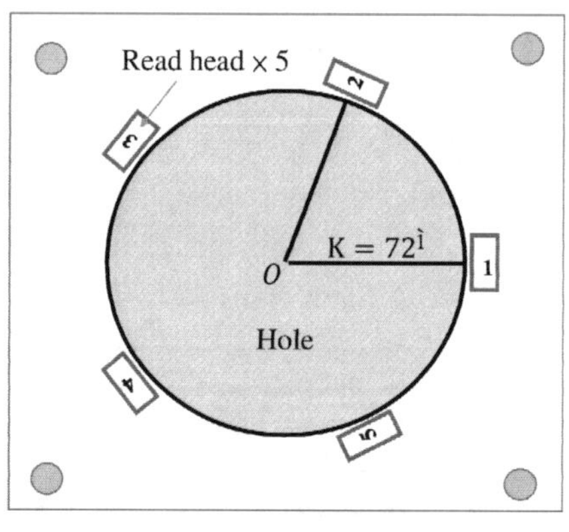

Figure 2. Five optical read heads mounted for PRAD

$$\alpha(\theta)= \frac{A_0(\theta)+A_1(\theta)+A_2(\theta)+A_3(\theta)+A_4(\theta)}{5} + \text{residual errors } (\theta) \tag{1}$$

$$\delta(\theta)= \beta(\theta)-\alpha(\theta) \tag{2}$$

3. Experimental setup and result

A Renishaw TONiC head is used as the optical read head, and the REXM ring scale has a 57-mm diameter. For the installation, the read heads were adjusted so that the signal strength around the full-axis rotation would achieve five green LEDs and not red flashing. Therefore, the installation is very easy and should only take a few minutes. The experimental result was compared the traditional calibration method. The used instruments of traditional calibration method for measuring the rotary angle were a polygon and autocollimator. A 12-face polygon was employed in this study. The comparison experiment setup is shown in Figure 3. For autocollimator specifications, the measuring range was from −2000″ to +2000″ and the maximum permission error was less than ±0.1″. For the polygon specifications, the angle accuracy was less than ±1″. We adjusted the tilt angle (y-axis direction) to less than 10″ at approximately 360° rotation and then start to calibrated angle of rotary angle of the actuator. The traditional method had high resolution and accuracy but there were some disadvantages. One, the optical alignment may be inconvenient for an engineer, and second, the data point is limited by the number of faces of the polygon.

The tested motor of actuator was an FHA-mini rotary actuator, which consists of a motor, harmonic gear, and encoder. The motor is a pancake brushless servo. The reduction ratio of the harmonic gear is 50:1. The encoder resolution is 6,553,600 count/rev in the absolute mode, such that a driven resolution of 0.197″ can be achieved. During testing, the motor rotation was set to 15 rpm to reduce the dynamic error influence. The resolution of PRAD is 0.036 arcsec. Finally, the comparison result for the motor one-way position error in the PRAD and autocollimator is shown in Figure 4.

Through equation (1), we analyzed residual errors for two tolerances of the shaft couplings. For the tolerance check, we used a dial indicator with 2-μm resolution to measure the runout. As shown in Figure 5, the difference in the two results of the runout is approximately a factor of 10. However, the two analyzed residual errors are close to each other. Hence, we know that the tolerance will not affect the measurement accuracy of PRAD.

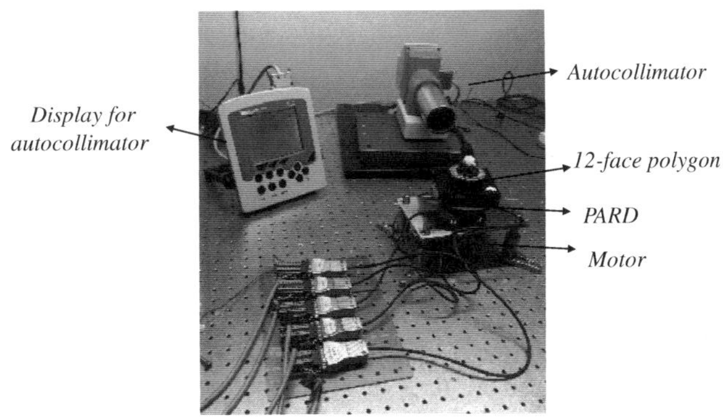

Figure 3. Experimental setup

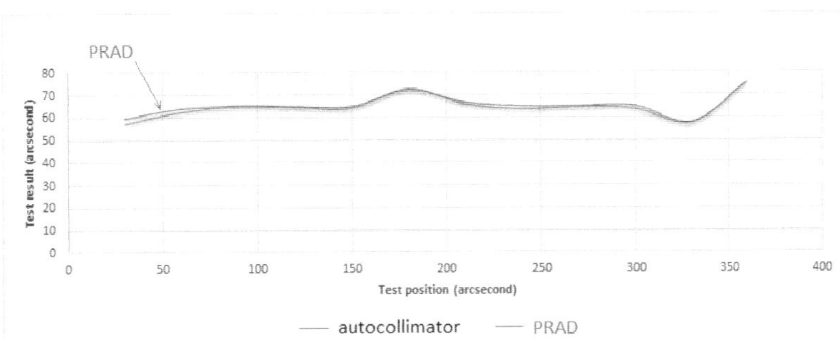

Figure 4. Comparison for motor one-way position error in PRAD and autocollimator

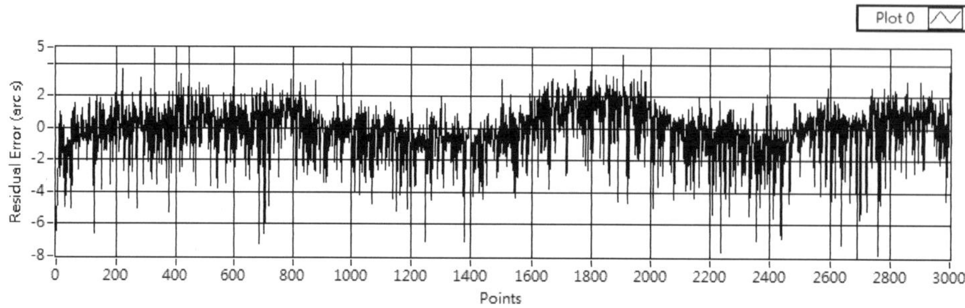

(a) poor shaft coupling

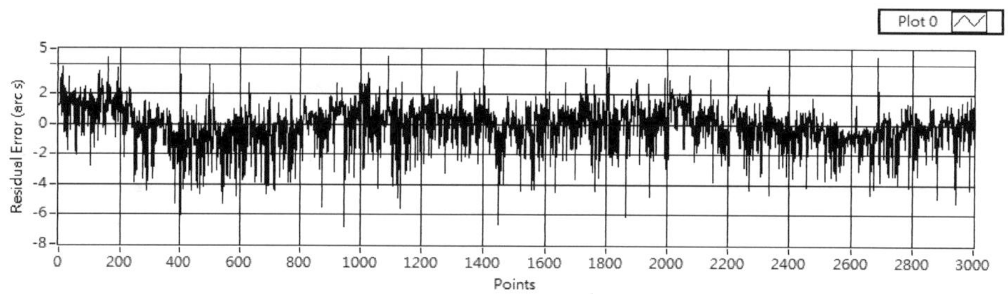

(b) better shaft coupling

Figure 5. Comparison of the residual error analysis for two shaft couplings

4. Discussion and conclusions

In this paper, PRAD was successfully developed to measure the angular position errors in the rotary actuator quickly. A traditional method including a 12-face polygon and an autocollimator are used to check and compare the PRAD performance. The experimental results showed that the structure of the PRAD is simple. However, the difference beyond 360° is only 0.7 arcsec. This small difference may arise from polygon specification tolerance and it is suitable for industry applications. Future work includes attempts to find more rotary geometric errors from the PRAD measuring results.

5. Reference

[1] ISO 230-1 2012 *International Organization for Standardization*. p13-14.
[2] Renishaw 2017 White paper *The Accuracy of Angle Encoders.*
[3] Watanabe T, Kon M, Nabeshima N and Taniguchi K 2014 *Meas. Sci. Technol.* **25**.
[4] Masuda T and Kajitani M 1993 *J. Robotics and Mechatronics* vol 5 p 448.
[5] Masuda T and Kajitani M 1989 *Precision Eng.* vol 11(2) p 95.
[6] Lou Z. F., Hao X. P., Cai Y. D., Lu T. F., Wang X. D. and Fan K.C. 2019 *Sensors* 2855.

Acknowledgments

The authors wish to acknowledge the financial support received from the Ministry of Economic Affairs in Taiwan. We also thank our supervisor, Dr. Fu, for assistance with the experiments.

IOP Conf. Series: Materials Science and Engineering **841** (2020) 012003 doi:10.1088/1757-899X/841/1/012003

Tests of Electrical Motor for Installation in the Wheel Hub of an Electric Car

B Bedkowski[1], P Dukalski[1], T Jarek[1] and T Wolnik[1]

[1] Łukasiewicz Research Network - Institute of Electrical Drives and Machines KOMEL, Katowice, Poland

E-mail: info@komel.com.pl

Abstract. The article presents the construction of test stand for testing an electric permanent magnet synchronous motor (PMSM) with an external rotor. The results of laboratory tests such as temperature rise tests for various motor parameters (such tests allow to determine temperatures in important motor components during different driving conditions - calculated for Fiat Panda III) and load tests with the determination of efficiency maps in the entire range have been presented. It was assumed that the car's drive is fitted with two motor in the rear axle.

1. Introduction

In recent years, the KOMEL Institute of Drives and Electrical Machines has carried out a number of design and research works in the field of drives for various types of electric vehicles [1-4]. One of the currently realized projects in the field of e-mobility concerns the direct drive motor installed in the wheel hubs of vehicles, which is related to the need to carry out research in wide range. Among the issues of construction and experimental works aimed at developing and manufacturing the product, it can be distinguished the most important ones: defining design assumptions, modeling and computer simulation, developing design documentation, making a prototype and experimental verification. The individual process steps are strictly interdependent. Particular importance for the assessment of the effectiveness of the design process are the results of the real model tests. They allow to assess the correctness of computational and design work as well as to verify initial assumptions.

As part of the project carried out at the KOMEL Institute, a PMSM motor with an external for installation in the wheel hub was designed and manufactured [5]. One of the prototypes is shown in Fig. 1 and Fig. 2. In order to determine the parameters of the motor (electromechanical and thermal), a number of tests were carried out to analyze the conformity of the real and design characteristics.

2. 3D model of the tested motor

The drive system equipped with two motors of this type should ensure dynamic driving of a Fiat Panda type car. For cars with a larger weight, in order to obtain sufficiently high driving parameters, it is preferable to use a 4-wheel drive system. The overall dimensions of the developed motor have been adapted to the 17 "rim. The available design space is limited by the outer diameter of the rim, its shape, rim shoulder (ET) and the brake drum (if included in the braking system). The presented motor has an external rotor. The rotating element is the body in which the magnetic core is installed with fixed permanent magnets. The fixed element is an anchor disc with a supporting structure, a labyrinth cooling system and a stator with windings. The rotor position sensor is necessary to control the motor.

Figure 1. A prototype of PMSM motor for installation in the wheel hub of an electric car

Figure 2. 3D model of the SMZs200S32 motor (prod. KOMEL) - cross-section

3. Description of the test bench

The motor tests were carried out on the prepared test stand, which is shown in Figure 3.

Tested motor was coupled through torque meter with an auxiliary DC machine. The tested object was powered from the Sevcon Gen4 Size8 controller dedicated to the drives of electric vehicles. During the tests, the inverter was supplied with a constant voltage of nominal value. As the necessary rotor angular position sensor, a magnetic absolute encoder with an analogue Sin-Cos output was used. Control of the motor controller was carried out by a control and measurement computer with a dedicated application.

For the measurement of electromechanical parameters, a high-performance power analyzer with a current transducer was used. Such a system enabled measurement of motor voltages and currents (and determination of active power and power factor), voltage and current consumed by the inverter (and determination of the power consumed by the entire electric drive) as well as mechanical quantities: torque and speed. On the basis of the obtained results for each static point, the efficiency of the motor, controller and the entire electric drive was determined. Archiving of measured parameters was carried out on a control and measurement computer using dedicated software. The tested motor, as a prototype object, was equipped with a series of resistive temperature sensors placed in strategic and most important places in terms of thermal analysis such as: stator windings (winding in slot and end winding) stator core and water cooling system jacket. The sensors have been connected to a multi-channel recorder enabling the archiving of temperature changes at individual points. Due to the fact that this motor is characterized by high values of the supply voltage frequency, it was decided to measure the temperature of permanent magnets placed in the external rotor.

Due to the fact that this motor is characterized by high values of the supply voltage frequency, it was decided to measure the temperature of permanent magnets placed in the external rotor [6]. For this purpose, a small wireless temperature logger was developed, which after installing on the surface of the rotor and connecting the sensor placed on the magnet enables continuous temperature recording. Block diagram of the recorder is shown in Fig. 4. The main element of the system is a microcontroller (μC) powered by lithium-ion battery (BAT) and voltage regulator (VR). The resistance of a small foil PT100 sensor placed on the magnet is converted to a digital by 15-bit RTD-to-digital converter (RTD-Conv) and goes to the microcontroller (μC) where it is converted into temperature. The temperature results are recorded at the specified time on the microSD card. The microcontroller used is a small size 32-bit module with ARM Cortex M0 + core. It is characterized by low power consumption with good technical parameters - for this reason it is often chosen for use in mobile electronic devices, among others according to the concept of the Internet of Things. The described measurement module together with the battery has been mounted on the part of the shaft. In order to limit the disturbances that could affect the system, a ferrite tape was used between the shaft and the recorder and all connections were made through shielded lines (in particular the RTD sensor lines). The use of a small Li-Ion battery with a capacity allows to perform long tests with measurement of the magnets temperature without charging.

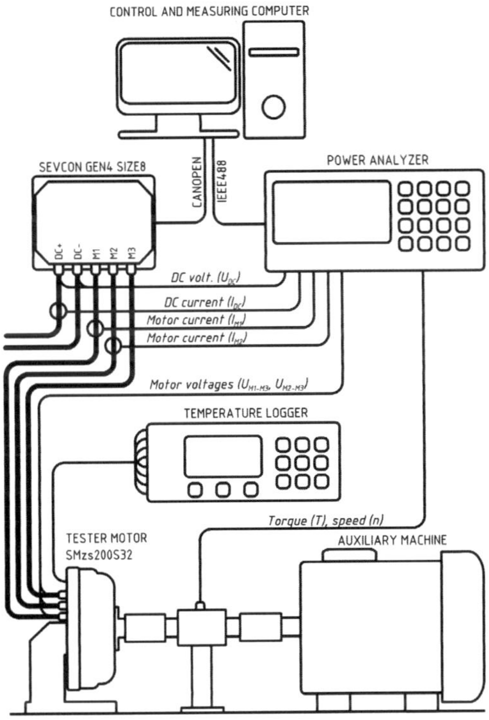

Figure 3. Schematic diagram of the test bench

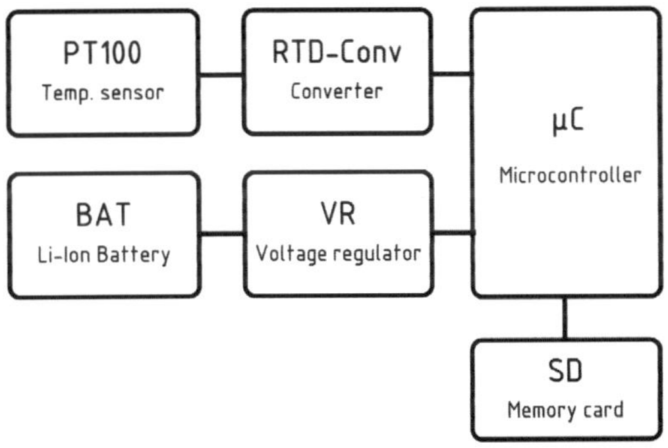

Figure 4. Diagram of temperature logger

4. Measurement results

In the case of motors intended for applications in the broadly understood electromobility, the significant parameter of the motor is its maximum torque and instantaneous overload (in order to ensure good dynamics of the vehicle). The motor should also be characterized by high efficiency in a wide range of speed [7].

Figure 5 shows the map of the efficiency of the motor made on the basis of the laboratory tests results. The map covers the full range of torque up to the maximum motor torque of 750 Nm and the full speed range up to 1400 rpm, which is equivalent to the vehicle speed (Fiat Panda, wheels with 17 "rims) equal to approx. 155 km/h. The following figures show the heating characteristics for various work points resulting from the actual driving parameters of the Fiat Panda.

The results of laboratory tests showed that the motor meets the requirements of the work parameters specified in table 1. With the speed of 1360 rpm, which is equivalent to the car's speed of 150 km/h, the motor is able to work for a long time with a much higher torque than it is required to overcome the motion resistance of the vehicle (Fig. 6). Such a torque reserve can be beneficial when driving in adverse weather conditions, e.g. in strong winds.

The operating temperature of the winding and permanent magnets are safe and are set at a level lower than the insulation class (180°C) and for permanent magnet during field weakening (120°C) [8].

Table 1 presents the calculated motor operation parameters for different driving conditions of the Fiat Panda with two motor built in the rear axle. Figures 7 and 8 show the temperature changes during operation under other conditions.

Table 1. Specification of motor operation parameters for various driving conditions

Case	P_m (kW)	T (Nm)	n (rpm)	Comments
1. The maximum speed of the vehicle on a flat surface - 150 km/h	21.4	150	1362	The calculated power is continuous
2. Vehicle speed 50 km/h - 20% elevation in the distance of 1 km	23.5	495	454	Time needed - about 1.5 minutes
3. Vehicle speed 70 km/h - 15% elevation in the distance of 5 km	26.3	395	635	Time needed - about 4.5 minutes
4. Vehicle speed 100 km/h - 10% elevation in the distance of 10 km	29.4	309	908	Time needed – about 6 minutes
5a. Acceleration of the vehicle 0-100 km/h	69.5	730	908	Acceleration in 10s
5b. Acceleration of the vehicle 0-100 km/h	59.0	620	908	Acceleration in 12s
5c. Acceleration of the vehicle 0-100 km/h	48.7	512	908	Acceleration in 15s

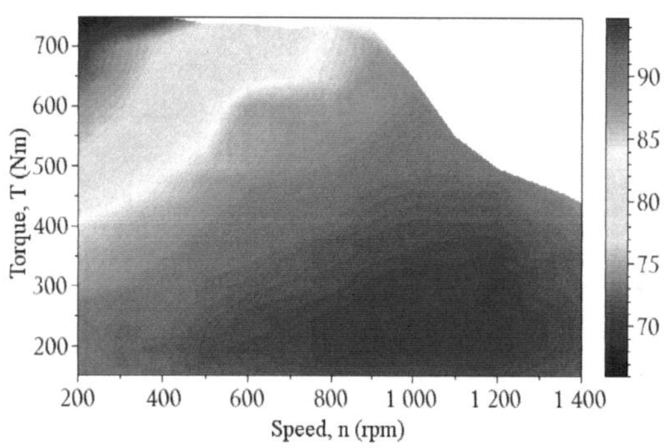

Figure 5. Motor efficiency map based on measurements

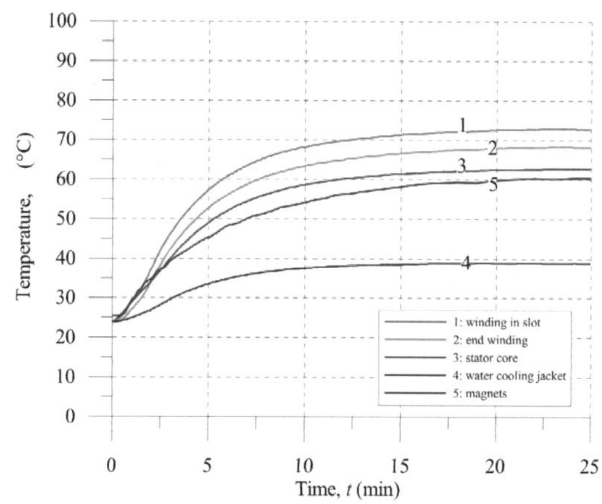

Figure 6. Temperature during the heating test: 1360 rpm, 150 Nm, 21.3 kW, 10 dm^3/min

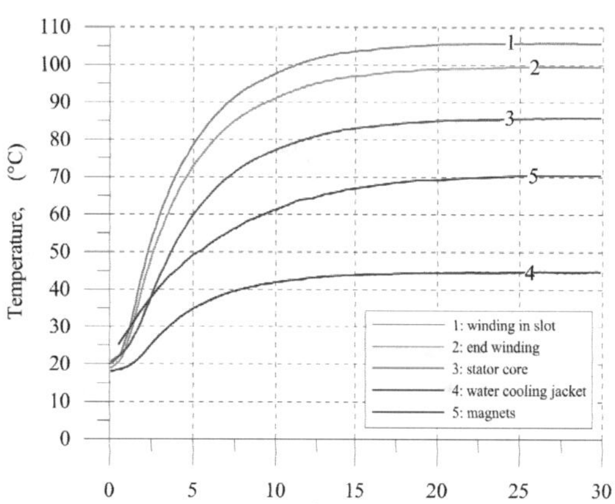

Figure 7. Temperature during the heating test: 950 rpm, 400 Nm, 40 kW, 10 dm^3/min

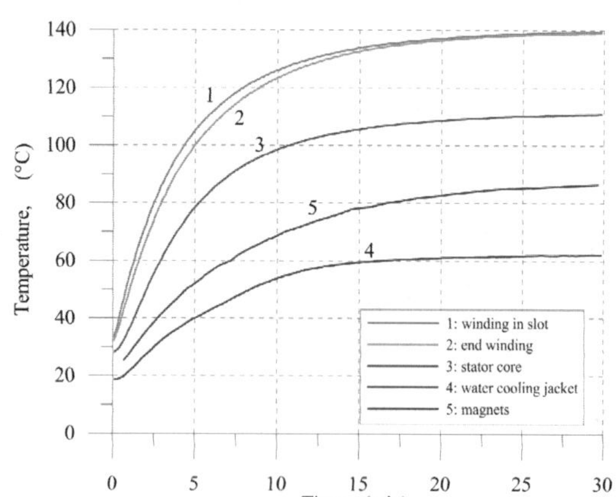

Figure 8. Temperature during the heating test: 1060 rpm, 450 Nm, 50 kW, 10 dm^3/min

5. Summary

The developed test stand allows for testing traction motor for installation in wheel hubs of cars in the full load and speed range. One of the most important tests are heating tests that allow to determine the time in which the motor can work with given parameters. Such tests allow to determine temperatures in important motor components during different driving conditions, e.g. when driving at maximum speed, during acceleration or when driving on a slope. The measurement of the temperature of permanent magnets is also important for developing motor for installation in the wheel hub. Neodymium magnets have an acceptable range of operating temperature which, if exceeded, may cause demagnetization of the magnets. For this reason, at certain work points, it is the temperature of permanent magnets that can limit the range of the motor's operation, not the temperature of the winding [9]. The design and testing of the electrical motor for mounting in the wheel hub should be complemented by research on the impact of the use of additional unsprung [10] mass on the comfort of driving and the operation of the suspension system. Testing on the vehicle should allow to determine the maximum mass of motor, which has a significant impact on its design and operating parameters.

6. References

[1] Krol E and Bernatt J 2018 Optimum selection of electrical motor for urban bus drive *Proc. of International Conference on Advanced Automotive Technology* (ICAT), Gwangju, Korea.

[2] Bedkowski B and Radwanski W 2012 Traction calculations of the sports and recreation vehicle UTV *Zeszyty problemowe - Maszyny Elektryczne.* **3** 105-108.

[3] Dukalski P, Rossa R, Dzikowski A and Kurytnik I P 2015 Design of PM motor dedicated to longwall shearer advance drive *2015 16th International Conference on Computational Problems of Electrical Engineering (CPEE)* 2015 16th International Conference on Computational Problems of Electrical Engineering (CPEE) (Lviv, Ukraine: IEEE) pp 24–31

[4] Gawron S Glinka T and Bernatt J 2018 Conception, Design and Production of Electrical Drives for Special Purpose *Machine Dynamics Research Politechnika Warszawska ISSN 2080-9948.*

[5] Dukalski P Bedkowski B Wolnik T Urbas A and Augustynek K 2017 Design assumptions of wheel hub motor for electric cars *Zeszyty problemowe - Maszyny Elektryczne.* **114** 263-272.

[6] Ding X, Liu J and Mi C 2011 Online temperature estimation of IPMSM permanent magnets in hybrid electric vehicles *2011 6th IEEE Conference on Industrial Electronics and Applications* 2011 6th IEEE Conference on Industrial Electronics and Applications (ICIEA) (Beijing, China: IEEE) pp 179–83

[7] Sakai K, Hattori T, Takahashi N, Arata M and Tajima T 2001 High efficiency and high performance motor for energy saving in systems *2001 IEEE Power Eng. Soc. Winter Meeting. Conference Proc.* 2001 Winter Meeting of the IEEE P. Eng. Soc. vol 3 1413–8

[8] Hendershot J R and Miller T J E 2010 *Design of brushless permanent-magnet machines* (Venice, Florida: Motor Design Books)

[9] Bedkowski B, Dukalski P, Jarek T and Wolnik T 2018 The Efficiency Analysis of Various Structural Solutions of the Wheel Motor Cooling Systems *2018 XIII International Conference on Electrical Machines (ICEM)* 2018 XIII International Conference on Electrical Machines (ICEM) (Alexandroupoli: IEEE) pp 995–1000

[10] Dukalski P, Bedkowski B, Parczewski K, Wnęk H, Urbaś A and Augustynek K 2018 Dynamics of the vehicle rear suspension system with electric motors mounted in wheels *Eksploat. Niezawodn. - Maint. Reliab.* **21** 125–36.

Acknowledgments

Work funded for science in the years 2017-2019 as a research project LIDER VII, No. LIDER/24/0082/L-7/15/NCBR/2016 carried out in the Institute of Electrical Machines and Drives KOMEL called "Innovative Solutions for Direct Drive of Electric Vehicles".

IOP Conf. Series: Materials Science and Engineering **841** (2020) 012004 doi:10.1088/1757-899X/841/1/012004

A Novel System of Primary Oscillations for Micromechanical Gyroscope

P Baranov[1], T Nesterenko[1], B B Duc[1], and L V Hao[1]

[1]National Research Tomsk Polytechnic University, 30, Lenin Ave., 634050, Tomsk, Russia

E-mail: bpf@tpu.ru

Abstract. The paper proposes a novel system of primary oscillations produced by the microelectromechanical (MEMS) gyroscope. This system is based on a synchronous detection of the current passing through the power capacitances of MEMS gyroscope. This detection method enables the excitation and stabilization of primary oscillations during two measuring cycles. The proposed system of primary oscillations also enhances the operation speed and the accuracy of frequency control, and reduces the ambient temperature effect on the stability of the scaling factor of the MEMS gyroscope.

1. Introduction

A microelectromechanical (MEMS) gyroscope is an integrated circuit unit which is placed onto a crystalline silicon substrate (silicon wafer) comprising different-type MEMS structures and electronic components necessary for measuring angular velocities [1-4].

Whatever configuration of such a system which uses the Coriolis effect is proposed, it should have a subsystem which generates the mechanical motion (primary oscillations) of the inertial mass in the microgyroscope (MG) [5-7]. The functioning principles of this subsystem greatly determine such important MG parameters as its readiness time after switching, conversion factors of angular velocities and their resistance to destabilizing factors, conversion errors [8-11].

Various types of microactuators are used to provide the motion of the MG inertial mass along the selected axis. To these types belong electrostatic vibratory drives in which the interaction between MEMS structures and electronics is performed via a single capacity transmitter or a differential capacity transmitter.

Traditionally, two groups of comb electrode structures are used to provide the motion of the inertial mass and stabilization of primary oscillations [12-15]. The first group includes driving electrodes of the microactuator, while the second comprises driving feedback electrodes which gather measurement information in order to control and stabilize the primary oscillations. The primary oscillation control is performed by the phase-locked looped system (PLLS) that results in long-term transition processes and reduces the operation speed. Changes in the ambient temperature cause changes in PLLS parameters, thereby leading to the decrease in the amplitude stability, frequency of primary oscillations and, as a consequence, the stability of the scaling factor of the MEMS gyroscope.

This work proposes a novel system of primary oscillations produced by the MEMS gyroscope and describes it by a method of synchronous detection of the current passing through the power capacitances of MEMS gyroscope. This detection method provides the excitation and stabilization of primary oscillations during two measuring cycles

IOP Conf. Series: Materials Science and Engineering **841** (2020) 012004 doi:10.1088/1757-899X/841/1/012004

2. System of primary oscillations

Figure 1 shows a schematic of the MG configuration and figure 2 contains the structural circuit of the excitation and stabilization system of MG primary oscillations.

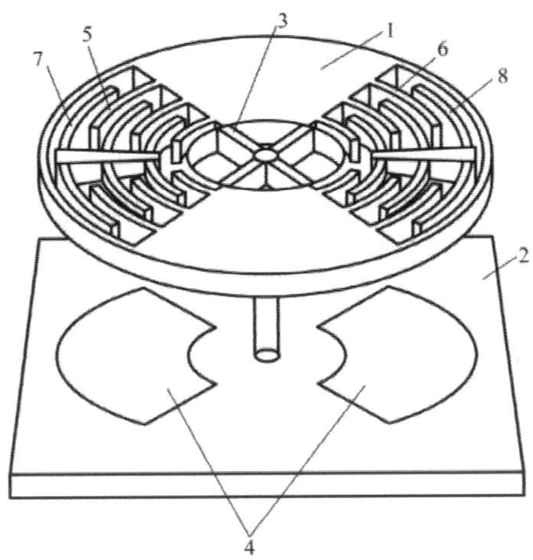

Figure 1. The MG configuration).

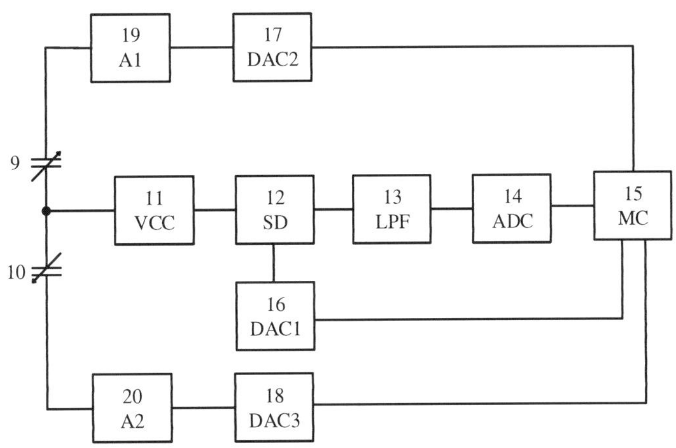

Figure 2. The structural circuit of the excitation and stabilization system).

The microgyroscope comprises the moving mass *1* which represents a disc movable relative to the support frame *2*. The moving mass is mounted to the support frame by spring elements *3*. Two ends of these spring elements are rigidly mounted to the moving mass *1*, whereas another two are mounted to the support frame *2*. Spring elements *3* are placed in the gyroscope such that to enable primary and secondary angular oscillations of the moving mass *1*. The former occur around *Z*-axis which is perpendicular to the plane of the moving mass *1*. The latter occur around *X*- and *Y*-axes within the plane of the moving mass *1*.

Stationary electrodes *4* are fixed to the support frame *2,* opposite to each other. Together with electrodes mounted to the bottom side of the moving mass, these electrodes form flat capacitors which are capacity transmitters of secondary angular oscillations of the moving mass *1* relative to the support frame *2*. The output of the capacity transmitter is connected to the gathering information unit.

Opposite to each other, movable comb electrodes *5, 6* are mounted to the moving mass *1,* which in combination with stationary comb electrodes *7, 8* form capacitors *9, 10* of the electrostatic comb drive. These stationary comb electrodes are shown in Figure 2.

The excitation system of MG primary oscillations operates as follows. The excitation and stabilization of MG primary oscillations are performed in two stages. The first stage includes setting the input parameters for the first digital-to-analog converter *16* (DAC 1) *via* the interface by the microcontroller *15* (MC). The following equation is used for setting these input parameters:

$$U_{r1}(t) = U_{rm}\sin(2\omega t), \tag{1}$$

where U_{rm} is the voltage amplitude at the output of the DAC *1*, V; ω is the voltage frequency, rad/s.

At the same time, the input parameters are set for the second digital-to-analog converter *17* (DAC 2) *via* the interface by the microcontroller *15* (MC). The following equation is used for setting these input parameters:

$$U_1(t) = U_m\sin(\omega t) + U_0. \tag{2}$$

The input parameters for the third digital-to-analog converter *18* (DAC 3) are obtained from the following equation:

$$U_2(t) = U_m\sin(\omega t + \pi) + U_0. \tag{3}$$

where U_m is the voltage amplitude, V; U_0 is the constant bias voltage, V.

The DAC 2 output voltage $U_1(t)$ is received by the input of the amplifier *19* (A1) and thus becomes stronger. The amplified output voltage $U_1(t)$ from A1 amplifier is received by the condenser *9* of the MG comb drive. The condenser *10* of the MG comb drive receives voltage from the DAC 3 output. This voltage is amplified by the second amplifier *20* (A2). As a result, the current flows through condensers *9* and *10* of the comb drive and then it flows to the input of the voltage/current converter *11* (VCC) and converts into the analog voltage. The output voltage from the voltage/current converter *11* is supplied to the first input of the synchronous detector *12* (SD), while the voltage from the first digital-to-analog converter *16* is supplied to its second input. The low-pass filter *13* (LPF) filtrates the voltage from the synchronous detector *12* which is then digitalized by the analog-to-digital converter *14* (ADC). The digitalized voltage U_p which is proportional to the inphase component of the total current flowing through the comb drive condensers *9* and *10*, delivers to and stores in the microcontroller *15*.

At the second stage, the output parameters of the digital-to-analog converter *16* are set by the microcontroller *15 via* the interface using the following equation:

$$U_{r2}(t) = U_{rm}\cos(2\omega t). \tag{4}$$

The procedures of the first stage are then repeated. The digitalized voltage U_q which is proportional to the quadrature component of the total current passing through the comb drive condensers *9* and *10*, delivers and stores in the microcontroller *15*.

The following equations are used by microcontroller *15* to compute the phase shift φ in the total current passing through the comb drive condensers *9* and *10:*

$$\varphi = \arctan\left(\frac{\lim\limits_{T\to\infty} 1/ET\int_0^T U_I(t)U_{r2}(t)dt}{\lim\limits_{T\to\infty} 1/ET\int_0^T U_I(t)U_{r1}(t)dt}\right), \tag{5}$$

where $U_I(t)$ is the output voltage from the voltage/current converter, V; E is the denominator of the synchronous detector, V; T is the time of integration, s.

When the phase shift is not equal to 90° which correspond to the maximum amplitude of MG primary oscillations, microcontroller *15* computes their resonance eigen-frequency as

$$\omega_r = \frac{\omega + \omega\sqrt{1+4Q^2\tan(\varphi)^2}}{2Q\tan(\varphi)}, \tag{6}$$

where Q is the quality factor of the primary oscillation channel of MG.

Next, microcontroller *15* changes the frequency ω *via* the interface at the output of the second digital-to-analog converter *17* and the third digital-to-analog converter *18* by the estimated value of the resonance eigen-frequency of MG primary oscillations.

Thus, the excitation and stabilization of primary oscillations around *Z*-axis occurs during two measuring cycles.

Under the rotation (angular velocity) of the support frame *2* around *X*- and *Y*-axes, Coriolis forces appear and enable the moving mass *1* to execute secondary angular oscillations around *X*- and *Y*-axes. Amplitudes of these oscillations are proportional to the measured angular velocities and convert into electric signals by capacity transmitters. The latter are formed by the stationary electrodes *4* fixed to the support frame *2* and movable comb electrodes fixed to the bottom side of the moving mass *1*.

3. Conclusions

The system of primary oscillations produced by the microelectromechanical gyroscope was proposed on the basis of synchronous detection of the inphase and quadrature components of the total current passing through the comb drive electrodes. Calculations of the resonance eigen-frequency produced by MG primary oscillations allowed us to excite and stabilize the primary oscillations at two measuring cycles. This was because the resonance eigen-frequency was mathematically calculated independently of changes in the ambient temperature, that increased the stability of the scaling factor of the MEMS gyroscope. One synchronous detector and one analog-to-digital converter were used to detect both inphase and quadrature components of the total current. The removal of the phase-locked looped system from the structural circuit of the excitation and stabilization system resulted in its simplification and the enhancement of the gyroscope reliability.

4. References

[1] Prikhodko I P, Zotov S A, Trusov A A and Shkel AM 2013 *J. Microelectromech. Syst.* **22** 1257
[2] Kajánek P 2014 *Geoinformatics FCE CTU.* 2014, **12** 61
[3] Cho S Y, Lee H K and Park C G 2013 *J. Dyn. Syst. Meas. Control.* **136** 1
[4] Robert O, Ann G D, and John L C, 2006 *MEMS and Microstructures in Aerospace Applications* (FL, USA: CRC Press) p. 4.
[5] Aaltonen L and Halonen K A I 2010 *Analog. Integr. Circuit Signal.* **63** 465
[6] Woon-Tahk Sung, Sangkyung Sung, June-Young Lee, Taesam Kang, Young Jae Lee and Jang Gyu Lee 2008 J. *Micromech. Microeng.* **18** 1
[7] Cui J, Chi X Z, Ding H T, Lin L T, Yang Z C and Yan G Z 2009 *J. Micromech. Microeng.* **12** 1
[8] Feng R, Qiu A P, Shi Q and Su Y 2011 *Adv. Mater. Res.* **403–408** 4237
[9] Xia D, Chen S, Wang S and Li H 2009 *Sensors* **9** 8349
[10] Yoona S W, Lee S and Najafia K 2012 *Sens. Actuators A: Phys.* **180** 32
[11] Yoona S W, Lee S and Najafia K 2011 *Sens. Actuators A: Phys.* **171** 163
[12] Yang B, Zhou B L and Wang S R 2006 *J. Phys. Conf. Series.* **34** 57
[13] Yang B, Li H S, Zhou B L and Wang S R 2009 *Sensors* **9** 3357
[14] Eminoglu B, Alper S E and Akin T 2011 *Procedia Eng.* **25** 1309
[15] Baranov P, Nesterenko T, Tsimbalist E and Vtorushin S 2017 *Meas. Sci. Technol.* **28** 064004

Acknowledgement

This work was performed in National Research Tomsk Polytechnic University. The authors like to acknowledge the financial support from the Ministry of Education and Science of the Russian Federation (Agreement N 14.578.21.0232, unique identifier RFMEFI57817X0232).

Experimental Study on Torsional Vibration of Engine Crankshaft and Cylinder Block

Pandeng Zhang[1, a], Zhao Liu[*1], Yutian Zhu[1], Jianming Zhuo[2], and Haijian Chen[2]

[1] School of Mechanical Engineering, Tongji University, Shanghai, China
[2] King Long Automotive Engineering Research Institute, Xiamen King Long United Automotive Industry Co., Ltd, Xiamen, Fujian, China

Email: zhangpandeng@tongji.edu.cn

Abstract. Engine cylinder block vibration usually does not consider the influence of torsional vibration of crankshaft. Because cylinder block and crankshaft vibrate separately, crankshaft torsional vibration may have a coupling effect on the vibration of the cylinder block, especially when the crankshaft torsional vibration is very large when engine is idling. In order to evaluate the coupling effect, a pilot test program was designed and developed under idle no-load condition. The torsional vibration of the crankshaft comes from the encoder sensor, and the torsional vibration of the cylinder block comes from the calculation of two acceleration sensors arranged on the cylinder block. This paper derives the formula for extracting the torsional vibration signal of the cylinder from two acceleration signals. Data analysis of spectrum and coherence reveal that the torsional of the crankshaft is greater than the cylinder block at the engine 3rd order frequency 35Hz, and the torsional vibration signals of the two have strong coupling at integer multiples of the 0.5th order frequency of the engine in the range of 100Hz that is prone to vibration problems. Through the above tests and data analysis, it is more accurate to consider the influence of crankshaft torsional vibration when studying engine vibration.

1. Introduction

Engine is the main vibration source of various mechanical products, and the vibration generated by it directly affects product reliability and comfort. Therefore, it is of great engineering significance to study the vibration of the engine. Engine vibration is transmitted to other parts through cylinder block. The noise generated by cylinder block vibration also accounts for an important part of engine noise. Therefore, the vibration of cylinder block is an important aspect of engine vibration. In general, when calculating the vibration response of the cylinder block, the coupling between the cylinder block and other parts is ignored. The load on the cylinder block is obtained by the method of multi-body dynamics, and response calculation and vibration prediction are carried out by using the finite element method, and weak parts of the cylinder block are found and modified to carry out the virtual design of low-noise engine [1-10].

The vibration generated by the crankshaft is three-dimensional vibration [11, 12]. Between the crankshaft and the engine block, the force can be transmitted through the combustible gas in the cylinder, and can be transmitted through the bearing of the crankshaft. Therefore, the possibility of

IOP Conf. Series: Materials Science and Engineering **841** (2020) 012005 doi:10.1088/1757-899X/841/1/012005

coupling between the two is very high. For many engines, subtle differences in the size of pulleys and flywheels can lead to dramatic changes in engine structure noise and vibration [13].

At present, rare documents about the torsional vibration coupling of cylinder block and crankshaft have been published. The coupling between the crankshaft and the drive system of the front-end accessory is studied only in some literature [14].

Considering the complexity of the theory, this article intends to do a test to find possible coupling relationship between crankshaft and cylinder block. As for test condition, the diesel engine's torsional vibration during idling is relatively large [15]. Therefore, idling condition was selected.

The torsional vibrations of the crankshaft and cylinder block of an engine were obtained by sensors of encoder, laser sensor and accelerometers, and the spectrum and coherence were analysed. The data show that the crankshaft torsional vibration is larger than the cylinder during idle speed, and the crankshaft and cylinder block torsional vibration are strongly coupled at integer multiples of the 0.5th order frequency of the engine. Therefore, it is more accurate to consider the influence of crankshaft torsional vibration when studying engine vibration.

2. Test Plan Design

In order to find the coupling relation through actual test data, the data have to be obtained through synchronous test in idle and no-load condition.

The crankshaft torsional vibration signal is achieved by the encoder mounted on the crankshaft belt pulley [16]. Vibration technology can provide useful information for diesel engine fault detection [17], and the torsional vibration of the cylinder block in this test is achieved by the appropriate calculation of the data of two acceleration sensors installed on the cylinder block. For phase reference, a laser sensor is arranged near the crankshaft belt pulley, that is, a reflective strip is pasted on the pulley. When the crankshaft turns one turn, the laser sensor will generate one impulse voltage signal, as shown in figure 1.

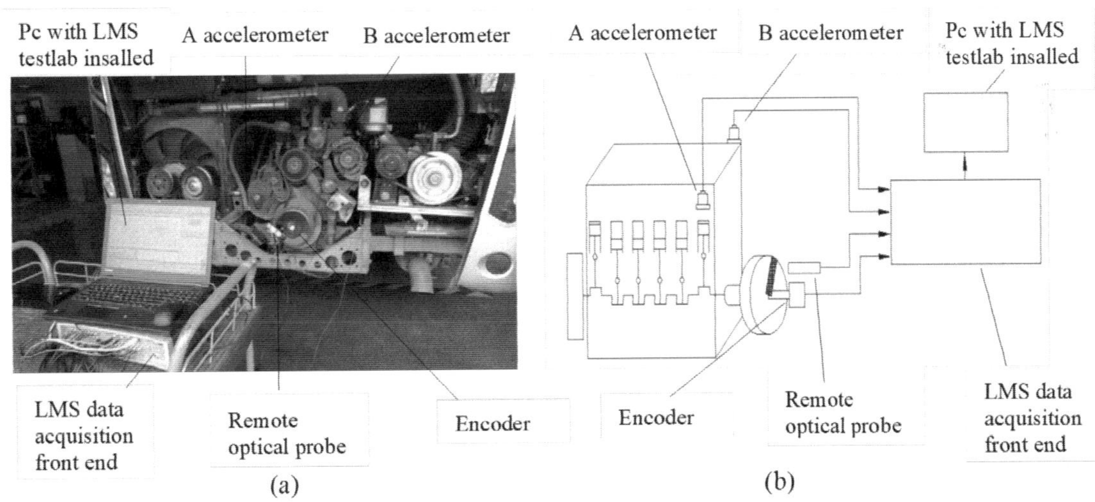

Figure 1. Testing sensors and measuring points.

According to the plane analysis, there are 3 degrees of freedom, namely, horizontal direction, represented by X, vertical direction, represented by Y, and rotation direction, represented by θ.

The angular acceleration signal of the cylinder block is calculated from the acceleration signals at two points, point A and point B. Schematic diagram of torsional vibration test and calculation is shown in figure 2.

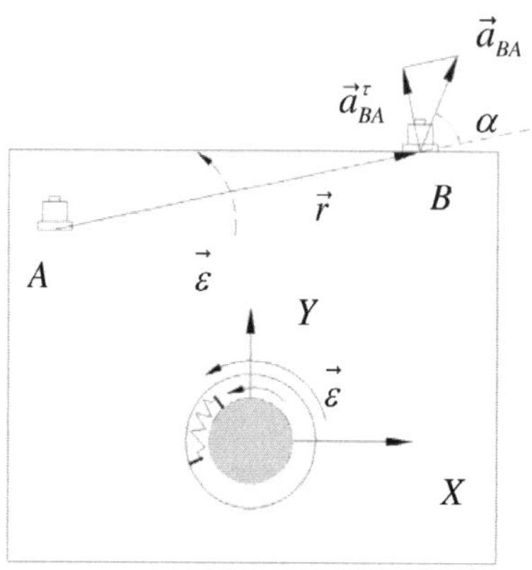

Figure 2. Schematic diagram of torsional vibration test and calculation.

$\vec{r}$ is the vector from point A to point B, and $\vec{\varepsilon}$ is the angular acceleration vector of the cylinder block, because of their vertical relationship, their norm of vector product is:

$$\left|\vec{\varepsilon}\times\vec{r}\right| = \left|\vec{\varepsilon}\right|\cdot\left|\vec{r}\right| \tag{1}$$

The magnitude of $\vec{\varepsilon}$ multiplied by the magnitude of $\vec{r}$ equals the magnitude of the tangent acceleration $\vec{a}_{BA}^{\ \tau}$:

$$\left|\vec{\varepsilon}\right|\cdot\left|\vec{r}\right| = \left|\vec{a}_{BA}^{\ \tau}\right| \tag{2}$$

In the above formula, $\vec{a}_{BA}$ refers to the relative acceleration from point A to point B, and $\vec{a}_{BA}^{\ \tau}$ is its tangential vector which can be obtained through the projection in the tangential direction of the acceleration $\vec{a}_{BA}$:

$$\vec{a}_{BA}^{\ \tau} = \vec{a}_{BA}\cdot\sin\alpha \tag{3}$$

Where, α refer to the angle between $\vec{a}_{BA}$ and $\vec{r}$.

The above expression can also be expressed as the vector product of $\vec{a}_{BA}$ with a unit vector $\frac{\vec{r}}{\left|\vec{r}\right|}$:

$$\vec{a}_{BA}^{\ \tau} = \frac{\vec{a}_{BA}\times\vec{r}}{\left|\vec{r}\right|} \tag{4}$$

Combine the above formulas, the following formula can be obtained:

$$\left|\varepsilon\right| = \frac{\left|\vec{a}_{BA}\times\vec{r}\right|}{\left|\vec{r}\right|^2} \tag{5}$$

$\vec{a}_{BA}$ and $\vec{r}$ are respectively defined as:

$$\vec{a}_{BA} = \vec{a}_B - \vec{a}_A = \left[(a_{Bx}-a_{Ax}),(a_{By}-a_{Ay})\right] \tag{6}$$

IOP Conf. Series: Materials Science and Engineering **841** (2020) 012005 doi:10.1088/1757-899X/841/1/012005

$$\vec{r}=\left[(r_{Bx}-r_{Ax}),(r_{By}-r_{Ay})\right] \tag{7}$$

Where, $\vec{a}_A$ and $\vec{a}_B$ refer to the acceleration vector of point A and point B respectively, a_{Ax} and a_{Ay} refer to X and Y direction of $\vec{a}_A$ respectively, a_{Bx} and a_{By} refer to X and Y direction of $\vec{a}_B$ respectively, r_{Ax} and r_{Ay} refer to the X and Y coordinates of point A respectively, and r_{Bx} and r_{By} refer to the X and Y of point B respectively.

The formula for calculating the calculated angular acceleration of the cylinder is as follows [18, 19]:

$$\ddot{\theta}=\frac{(a_{Bx}-a_{Ax})(r_{By}-r_{Ay})-(a_{By}-a_{Ay})(r_{Bx}-r_{Ax})}{(r_{Bx}-r_{Ax})^2+(r_{By}-r_{Ay})^2} \tag{8}$$

The cylinder block vibrates on three degrees of freedom in the plane, X, Y and θ. The acceleration sensor is consolidated with the cylinder block, so the measured data contains the part caused by the torsional vibration of the cylinder block and the part caused by the overall vibration signal in the X and Y direction. What we need is to extract the torsional vibration signal. By the calculation of two acceleration differences, signals only due to vibrations in the X and Y directions are eliminated, and the torsional vibration signals are extracted, as shown in equation (8). At the same time, the encoder does not collect vibrations in X and Y directions, so the torsional vibration of the crankshaft and cylinder block is not affected by the overall vibration of the engine in the X and Y directions.

The test object is an engine mounted on the passenger car, and the external accessory belt connected to the engine is completely disconnected, that is, the engine is in independent operation.

The relevant parameters of the engine and test instruments are given in table 1.

Table 1. Main parameters of the engine and sensors.

Engine/sensors	Parameter(units)	Value
Engine	Type of fuel	Diesel
	Cylinder arrangement	in-line
	Number of strokes	4
	Number of cylinders × bore diameter (mm) × stroke(mm)	6×105×130
	Piston displacement (L)	6.75L
	Compression ratio	18:1
	Rated power (kW) / rotational speed (r/min)	180/2300
Encoder	Pulses per revolution， PPR	100
	Maximum rotational speed(r/min)	18000
	Maximum Frequency $f_{\max 1}$ （kHz）	100
Laser sensor	Measuring range (r/min)	1-250,000
	Signal output	TTL(same as the input voltage)

IOP Conf. Series: Materials Science and Engineering **841** (2020) 012005 doi:10.1088/1757-899X/841/1/012005

Data acquisition front end	Pulse frequency of maximum speed f_{max2} (kHz)	40
	DSP (bit)	24
	Sampling rate per channel(kHz)	102.4
	Signal noise ratio(dB)	105
	Total harmonic distortion(dB)	<-100

When PPR value of the encoder is large and the maximum speed pulse frequency of the data acquisition front end is low, the speed may exceed the maximum allowable speed of the entire system. The maximum speed N_{max} by the combination of the encoder and the data acquisition front end is calculated according to formula (9) [20]. After calculation, the test speed did not exceed the value of N_{max} .

$$N_{max} = 60\min(f_{max1}, f_{max2})/PPR \tag{9}$$

The sampling frequency was 20480Hz, the frequency resolution was 0.5 Hz, and the window function is hanning window. The time domain data obtained by measurement and calculation are shown in figure 3.

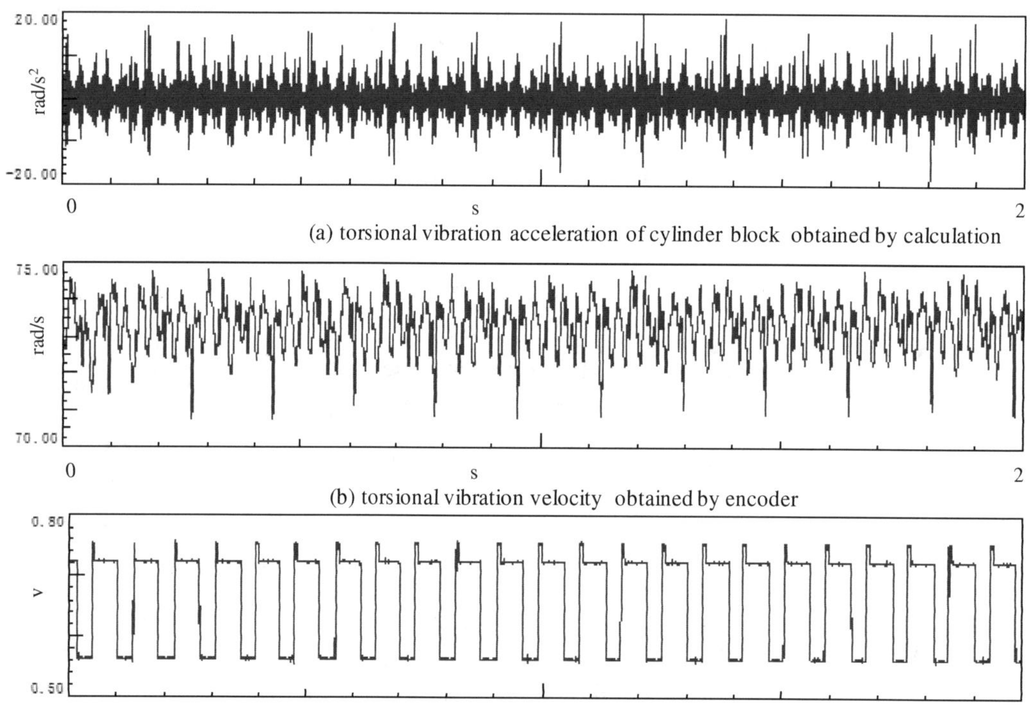

(a) torsional vibration acceleration of cylinder block obtained by calculation

(b) torsional vibration velocity obtained by encoder

(c) voltage signal produced by laser sensor

Figure 3. The time signals obtained through testing and calculation.

3. Test Results and Analysis

When conducting spectrum analysis, the vibration was expressed in the form of vibration speed. With the laser sensor as the reference signal, the obtained time-domain signal can be used to calculate the torsional vibration spectrum of the cylinder block and crankshaft, which are shown in the bode diagram in figure 4.

As can be seen from the figure 4, the torsional vibration of the crankshaft and the cylinder block at the idle speed of 700rpm is mainly reflected in the 35Hz of the third order of the engine. Crankshaft torsional vibration amplitude and phase respectively are 0.7 rad/s, 121°, cylinder block torsional vibration amplitude and phase are 0.059 rad/s, 153 °. It can be seen that the crankshaft torsional vibration is larger than the cylinder block, and the amplitude ratio is about 11.9.

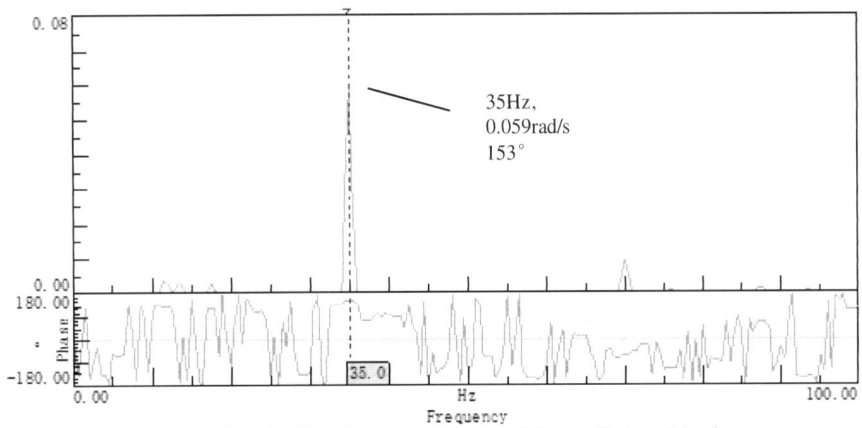

(a) torsional vibration spectrum of the cylinder block

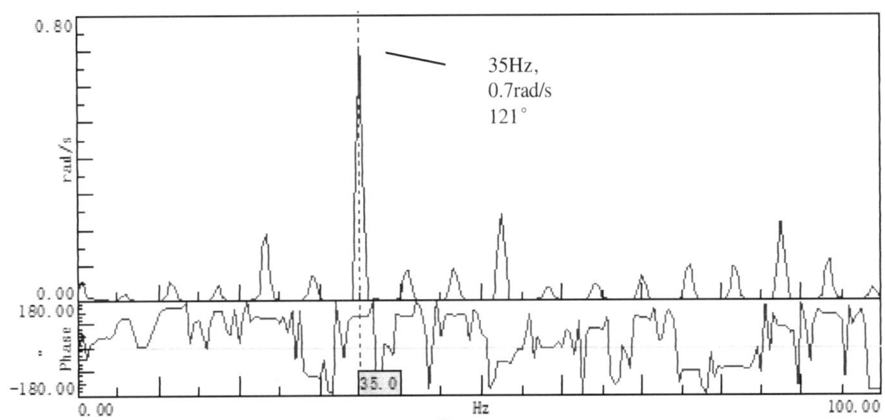

(b) torsional vibration spectrum of the crankshaft

Figure 4. Torsional vibration spectrum of the cylinder block and crankshaft in idle condition.

The coherence is utilized to detect the correlation of two signals in frequency domain [21, 22]. The definition of the (squared) ordinary coherence of signal X_i and signal X_j is shown in equation (10), where $\overline{S}_{ij}$ is the average crosspower, $\overline{S}_{ii}$ and $\overline{S}_{jj}$ are the average autopowers. The range of coherence is between 0 and 1, and a high value (close to 1) indicates that the output is almost entirely due to the input.

$$\gamma_{ij}^2 = \frac{\left|\overline{S}_{ij}\right|^2}{\overline{S}_{ii} \times \overline{S}_{jj}} \tag{10}$$

The result of the coherence analysis of the torsional vibration signal of the encoder and the torsional vibration signal of the cylinder block is shown in figure 5. At 700rpm, half order frequency is 5.84 Hz. As can be seen from figure 5, in the range of 100Hz that is prone to vibration problems, the coherence coefficients of half order frequency are all within the range of 0.9~1, while other frequencies are significantly lower. This indicates that the two signals have a strong internal connection at half order frequency, that is, the crankshaft has a strong coupling relation with the cylinder block. Therefore, if the coupling effect is not considered, there will be a certain error in the torsional vibration of the cylinder block.

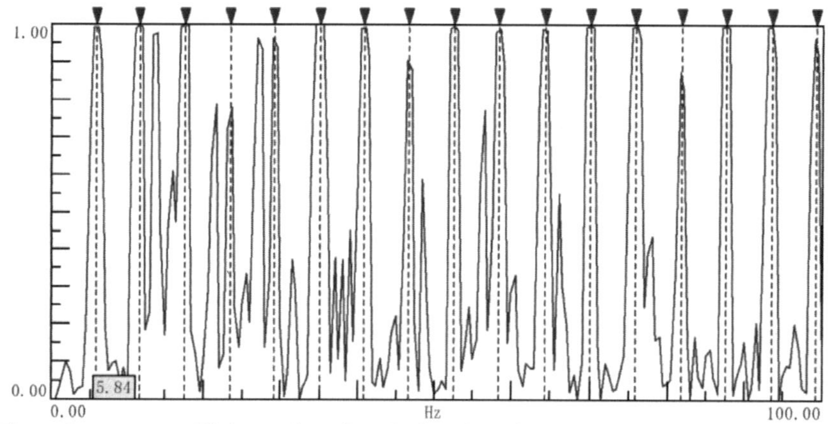

Figure 5. The coherence coefficient of torsional vibration signal of cylinder block and crankshaft.

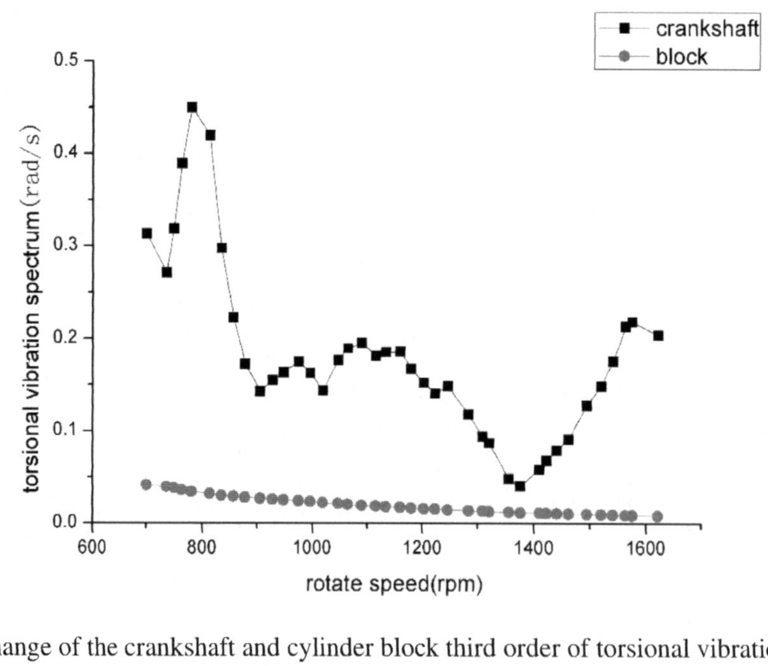

Figure 6. Change of the crankshaft and cylinder block third order of torsional vibration spectrum when the speed increases

Figure 6 shows the third order of torsional vibration spectrum within a commonly used speed range of 700-1600rpm. It can be seen that, when the speed increases, the torsional vibration of the engine crankshaft and the cylinder block changes, in which the torsional vibration of the cylinder block steadily decreases, while the torsional vibration of the crankshaft has a tendency to decrease while a larger fluctuation occurs.

4. Summary

Taking the signal of laser sensor as reference signal, the torsional vibration of an engine's crankshaft and cylinder block was tested and calculated in two working conditions of idle and no-load condition and rising speed condition, and the spectrum and coherence of the two signals were analyzed. The research conclusions are as follows:

- The torsional vibration of the crankshaft and cylinder block is mainly manifested in the 3rd order 35Hz, and under this condition the crankshaft torsional vibration is larger than the cylinder block, and the amplitude ratio is about 11.9.
- In the range of 100Hz that is prone to vibration problems, the coherence coefficients of integer multiple of the engine's half-order are all within the range of 0.9~1, while other frequencies are significantly lower. This indicates the crankshaft has a strong coupling relation with the cylinder block.
- According to the above analysis, it is more accurate to consider the influence of crankshaft torsional vibration when studying engine vibration.

5. References

[1] Y. Xingzhuo, F. Xing, ZhouYu, Prediction of the Acoustic Field of Box Type Vibrating Bodies by FEM—BEM, *Journal of Huazhong University of Science and Technology*, 21 (1993).
[2] J. Xianding, X. Lijuan, Ship Vibration, Shanghai jiaotong university press, 2011.
[3] R.J. Tyrrell, D.M. Croker, Engine Noise Practicalities and Prediction, Part 2 - Finite Element Analysis, *International Journal of Modern Physics* A, 14 (1987) 3687-3697.
[4] K. Maekawa, S. Morita, Calculation of radiated noise from cylinder block using finite element model, *Jsae Review*, 6 (1985) 228-239.
[5] Q. Lin, H.Z. Yong, J.W. Xin, L. Hong, Prediction method of radiated noise by engine block, *Journal of Jiangsu University*, (2008).
[6] Z. Jinling, S. Chunli, L. Yucheng, L. Kang, The Computer Simulation Analysis of Engine Structure Noise, Automobile Technology, (2006) 13-16.
[7] M. Jie, H. Zhiyong, Coupled optimization of internal combustion engine radiated noise and interior noise, *JOURNAL OF VIBRATION AND SHOCK*, (2015) 171-176.
[8] Y. Jianguo, Z. Xiaoming, L. Songhua, Vibration Response Calculation and Measurement of a New Medium-speed Diesel Engine, Diesel Engine, 30 (2008) 37-40+49.
[9] L. Czerny, M. Schwaderlapp, T. Wagner, NVH Optimization of a 16-Cylinder Diesel Engine, *Journal of the Electrochemical Society*, 147 (1993) 936-944.
[10] T. Kodama, Y. Honda, A Study on the Modeling and Dynamic Characteristics of the Viscous Damper Silicone Fluid by Using Vibration Control of Engine Crankshaft System, International *Journal of Mechanical Engineering and Robotics Research*, 7 (2016) 273-278.
[11] H. Okamura, A. Shinno, T. Yamanaka, A. Suzuki, K. Sogabe, Simple Modeling and Analysis for Crankshaft Three-Dimensional Vibrations, Part 1: Background and Application to Free Vibrations, *Journal of Vibration & Acoustics*, 117 (1995).
[12] T. Morita, H. Okamura, Simple Modeling and Analysis for Crankshaft Three-Dimensional Vibrations, Part 2: Application to an Operating Engine Crankshaft, *Journal of Vibration & Acoustics*, 117 (1995).
[13] H. Okamura, T. Morita, Influence of Crankshaft-Pulley Dimensions on Crankshaft Vibrations and Engine-Structure Noise and Vibrations, in: Noise & Vibration Conference & Exposition, 1993.

[14] W. Xiaoli, S. Wenbin, Z. Shaofei, C. Chao, Analysis of crankshaft torsional vibration using a coupled model of engine crankshaft systems and engine front end accessory drive systems, Journal of Vibration Engineering, 24 (2011) 505-513.
[15] Z. Dong, Z. Liangliang, G. wenzhi, C. wei, W. Jinli, L. Kai, Measurement and analysis of torsional vibration for a 2.8T diesel engine, *JOURAL OF MACHINE DESIGN*, 31 (2014) 73-76.
[16] S.D. Yu, X. Zhang, A data processing method for determining instantaneous angular speed and acceleration of crankshaft in an aircraft engine-propeller system using a magnetic encoder, Mech Syst Signal Pr, 24 (2010) 1032-1048.
[17] M.P. Robillard, W. Maalej, R.J. Walker, T. Zimmermann, An Approach on Fault Detection in Diesel Engine by Using Symmetrical Polar Coordinates and Image Recognition, Adv Mech Eng, 2014 (2014) 1-9.
[18] L. Zhao, Q. Xianrong, Z. Chuanming, Mechanical Vibration, Tongji University Press, 2016.
[19] Y. Yan, Advanced Engineering Mathematics, Huazhong University of Science & Technology Press, 2014.
[20] T. Xiangjun, Beginning NVH: Introduction and Advanced Tutorial on Noise, Vibration & Modal Analysis, 1 ed., CHINA MACHINE PRESS, 2018.
[21] L. Xingcai, H. Zhen, S. Gequn, Investigation on Mechanism of Axial Vibration for Internal Combustion Engine Crankshaft Using Coherent Analysis, Chinese Internal Combustion Engine Engineering, 25 (2004) 23-26.
[22] M.F. Dimentberg, D.V. Iourtchenko, A. Naess, Coherence function of transverse random vibrations of a rotating shaft, *J Sound Vib*, 292 (2006) 983-986.

Acknowledgment
The authors gratefully acknowledge the financial support provided by the Xiamen Science and Technology Bureau Project (3502Z20151002).

IOP Conf. Series: Materials Science and Engineering **841** (2020) 012006 doi:10.1088/1757-899X/841/1/012006

Fault Diagnosis of Ball Bearing Using Hilbert Huang Transform and LASSO Feature Ranking Technique

H T Thakker [1], V Dave [1], V Vakharia[1] and S Singh[2]

[1] Department of Mechanical Engineering, School of Technology, Pandit Deendayal Petroleum University, Gandhinagar-382007, Gujarat, India

[2] Department of Mechanical Engineering, Guru Nanak Dev University, Amritsar- 143005, Punjab, India

Email: harshit041196@gmail.com

Abstract. Bearings are one of the crucial components of any machine having rotary parts. They are employed to support and ensure smooth operations of the shafts in the rotary machinery. Therefore, any fault in the bearings can lead to a decline in the level of production and equipment. For this reason, it is important to monitor the bearing health. This paper presents a signal analysis technique for machine health monitoring using the Hilbert-Huang Transform (HHT). HHT is a time domain approach which extracts instantaneous frequency data from a signal by decomposing the signal into Intrinsic Mode Functions (IMF) using the Empirical Mode Decomposition (EMD). The Least Absolute Shrinkage and Selection Operator (LASSO) is used as feature ranking method which is used to improve the prediction accuracy by reducing input data to machine learning model by aiding to select only a subset of the feature vector rather than using all of the features. In the present work, training and tenfold cross-validation accuracy or two classifiers have been compared. The comparative analysis presented in this paper reveals that the utilization of LASSO as a feature ranking method shows a substantial decrease in the data to be handled and improving the diagnosis accuracy.

Keywords: Fault Diagnosis, HHT, LASSO, Machine Learning

1. Introduction

Bearings are one of the crucial components of any rotary machine, so it is recommended to identify a faulty bearing as soon as possible and take preventive actions as the faulty bearings can lead to catastrophic failure of the machine which leads to stoppage of production as well as can cause human casualties. The typical failures that occur in a roller bearing are due to localized defects in the inner-race, outer-race, or rolling elements. Due to the presence of such defects, a series of abnormal vibrations can be detected when the rolling elements interact with the surface of the defect and due to the intrinsic characteristics of each fault type, the frequency of the abnormal vibrations occurring are different depending on the type of fault [1]. Several techniques for detection of faults in rolling element bearing exist today. The vibration-based signal processing method is among one of the popular methods for diagnosing bearing faults. Fourier Transforms have been extensively used as a signal analysis tool for fault detection in bearings. However, the Fourier Transform is best suited for stationary signals i.e. signals having time-invariant properties [2]. The vibration signals that researchers encounter in faulty rolling element bearings are mainly non-stationary with a split second between changes in the amplitude of vibration signal which makes it difficult to quantify. Hence, it is crucial to detect these non-stationary signals effectively in order to prevent imminent machine failures. Wavelet Transform (WT) is an alternative approach to analyze non-stationary signals by utilizing the change in energy distribution with respect to time along with each frequency band [3]. One of the major drawbacks of the wavelet transform is the selection of base function for mother wavelet. In addition, WT faces a problem of energy leakage due to the limitation of the length of the wavelet function. To tackle these drawbacks of the WT, researchers have used the Hilbert-Huang Transform (HHT). HHT employs an adaptive time-frequency analysis which eliminates the requirement of prior functional basis [4]. Rather, HHT adaptively derives the basic functions from the data by utilizing Empirical Mode Decomposition (EMD). The derived basic functions obtain after performing on the original signals gives Intrinsic Mode Functions (IMF). A phenomenon known as Mode aliasing is observed in EMD and to overcome Ensemble Empirical Mode Decompositions (EEMD) was derived. Feature selection also known as feature ranking is gaining attraction now days. Feature ranking is a technique to select relevant features to avoid the curse of dimensionality as well to enhance the generalization capability of classifier used. In feature ranking technique the accuracy of machine learning model is evaluated using the ranked feature set [5-6]. The use of LASSO as a feature ranking and selection method provides benefits over the previous approaches in terms of the amount of data that needs to be handled when classifying the bearing faults. In order to rank the features, LASSO utilizes the linear dependency between the input features and the output values [7]. The features selected based on the LASSO feature ranking method are useful for improving the performance of the fault identification system.Fig.1.represent the methodology used for fault diagnosis.

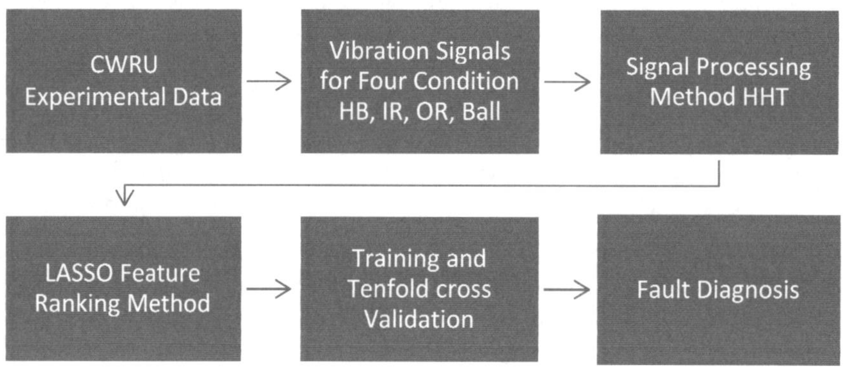

Figure 1. Flowchart for fault diagnosis using HHT

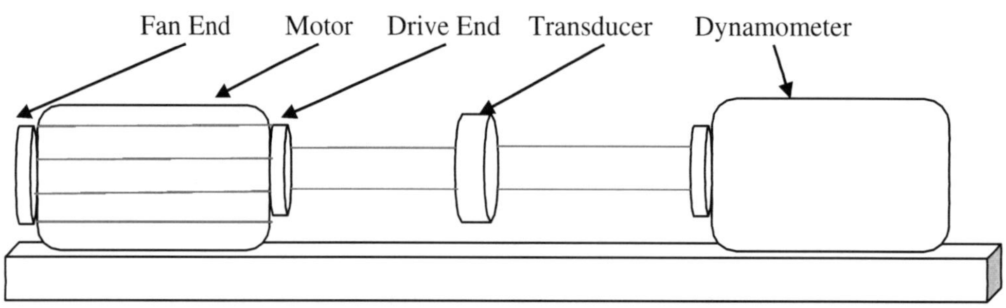

Figure 2. Bearing Test Rig

2. Experimentation

To conduct the study, Case Western Reserve University (CWRU) bearing data with different rotating speed and fault conditions are considered in the present paper. The schematic diagram of experimental set up with accelerometer mounted on the bearing housing at the drive end of the induction motor, which is connected to a torque transducer, and coupled with dynamometer, is shown in Fig. 2 Signals are captured from the drive end of bearing with the different fault conditions as: (1) healthy bearing

(HB); (2) inner race fault (IR); (3) ball fault (BF); (4) outer race fault (OR) and variation in rpm of shaft are 1730, 1750, 1772, and 1797 rpm respectively [8]. A 6205-2RSL JEM SKF deep-groove ball bearing is used for capturing signals from accelerometers, whose dimensions and specifications are referred from [9].

3. Hilbert-Huang transform

Due to the non-stationary nature of ball bearing in operating condition, Fourier Transform is not considered as an effective technique for fault diagnosis. According to the limitation of Heisenberg-Gabor inequality, the resolution of the signal in the time domain and frequency domain simultaneously cannot be done with conventional signal processing techniques. To overcome these limitations time-frequency analysis method known as the Hilbert-Huang Transform (HHT) is used [4]. This method employs Empirical Mode Decomposition (EMD) on the non-stationary and non-linear signals into various IMFs. The decomposed signal has different components known as Intrinsic Mode Functions (IMF) which captures the instantaneous frequency data as time-dependent function.

4. Feature selection using LASSO

In order to reduce the amount of data to be handled and enhance the performance of the classification models, we can reduce the number features from the feature vector by checking the relationships of the features with the class variables. Feature selection methods provide various benefits such as less computation time, maximizing prediction accuracy, and selection of relevant features from the signals captured after conduction experiments. Robert Tibshirani [10] introduced the method of Least Absolute Shrinkage and Selection Operator (LASSO). The method uses a shrinking process also known as regularization to penalize the coefficients of variables. The variables also known as feature, in present study, which have nonzero coefficient after the application of shrinking process, are selected as a relevant feature. When numbers of features are more and numbers of experiments are less then LASSO is useful, since the variance is reduced without a substantial increase in the biasedness [11]. In this way overfitting problem is also reduced due to selection of relevant features. Tibshirani used the quadratic program (QP) for least square regressions in LASSO. Table 1 shows the features ranked in descending order as selected by LASSO.

5. Fault identification techniques

After selecting a relevant feature from the LASSO-based feature ranking technique, the accuracy of the methodology used is assessed with two different classifier using machine learning techniques. Support Vector Machine (SVM) is a machine learning model used for classification as well as regression problems. To maximize the margin, the SVM regression algorithm works. Margin is defined as the distance between the separating hyperplane and training samples which are close to the hyperplane known as support vectors [12]. In this study, since the feature vector is extracted from ball bearings so non-linear kernel function PUK is used to fit the maximum-margin hyperplane so as to accurately predict the class of the input feature set. Random Forest employs the ensemble learning method which comprises of a combination of single-level decision trees and arrives at decision-based on the voting of a random set of available decision trees. For instance, in the initial phase, two third samples are considered for training and the remaining one third is considered for validation. The process is repeated ten times if ten-fold cross validation is performed on classifier [2]. In the process of model building each decision tree cast a vote for one class, and finally the calculated votes are used to assess the generalization capability of the classifier.

6. Results and discussion

In the present work, training and tenfold cross-validation using SVM and Random Forest, has been done. The relevant feature according to LASSO based feature ranking is shown in Table 1. In tenfold cross-validation process, ten iterations are performed when the data set is divided into ten-sized folds in such a way that nine parts are used for training and remaining part used for testing. Total of 64 instances and 20 features are used for the study. The variation of training and tenfold cross-validation accuracies with respect to the number of features using SVM and Random Forest can be observed in

Fig.3.With the first eight features selected by LASSO feature ranking method 100% and 85.93% training and tenfold cross-validation accuracy respectively is achieved while using the SVM model. For the Random Forest model, first nineteen features were selected using the LASSO Feature ranking method and 100% and 96.87% training and tenfold cross-validation accuracy respectively was achieved which diagnose the fault with considerable accuracy. Table 2 shows the result obtained through training and tenfold cross-validation. It is observed that Random Forest gives better result as compared to SVM.

Table 1. Selected Features Using LASSO

Sr.no	Feature Name	Sr.no	Feature Name
1	L Factor	11	Log Entropy
2	Median	12	Form Factor
3	Standard Deviation	13	Shannon Entropy
4	Mode	14	Shape Factor
5	Peak to RMS	15	Norm Entropy
6	Variance	16	RMS
7	Peak to Peak	17	Kurtosis
8	Signal to Noise and Distortion Ratio	18	Crest Factor
9	Root Sum of Square	19	Skewness
10	Sure Entropy	20	Mean

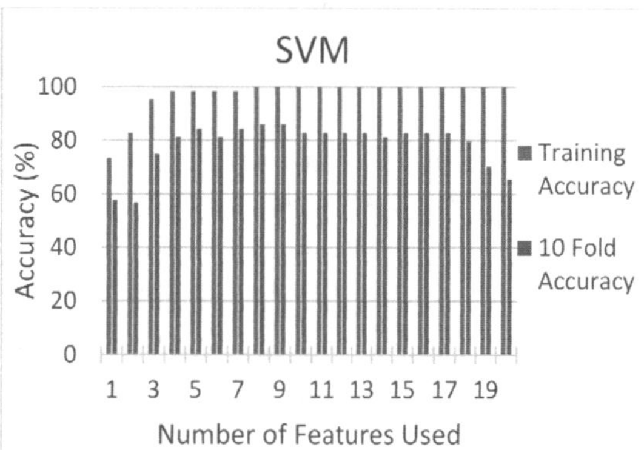

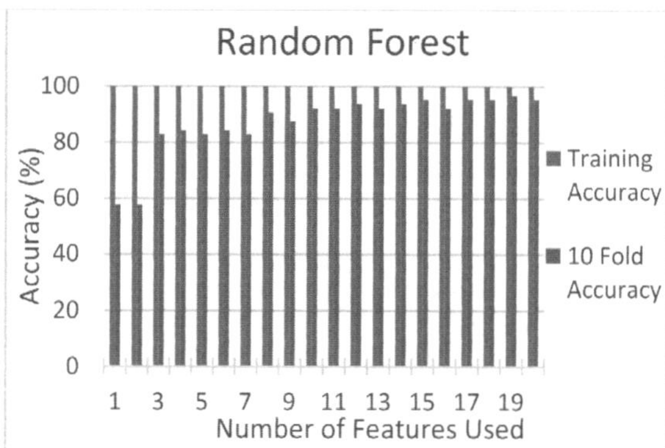

(a) (b)

Figure 3. Training and Tenfold Accuracy using SVM and Random Forest

Table 2. Training and Tenfold Cross Validation Accuracy

Sr. No	Classifier Name	No. of Features	Training Accuracy	Tenfold Accuracy
1.	Support Vector Machine	8	100	85.93
2.	Random Forest	19	100	96.87

7. Conclusion

In this paper, authors utilized a methodology to identify faults in bearings using HHT based signal processing technique, LASSO as a feature ranking method, and machine learning classification techniques. Twenty features are extracted through the HHT and then feature are ranked according to LASSO. The utility of features selected by LASSO feature ranking method provides a significant reduction in the number of features required for fault identification with satisfactory accuracy in diagnosing bearing faults. To diagnose the type of faults SVM and Random forest machine learning techniques used. Training and ten-fold cross validation performed on feature vector and it is observed that SVM and Random Forest classifiers gives 100 % training accuracy to diagnose bearing faults with both SVM and Random Forest. Similarly ten-fold cross validation accuracy of 85.93% and 96.87 % respectively achieved by utilizing eight and nineteen features respectively selected as per LASSO feature ranking method. The methodology used can be further compared with other signal processing methods and feature ranking techniques.

8. References

[1] Tang G, Pang B, Tian T, Zhou C. Fault diagnosis of rolling bearings based on improved fast spectral correlation and optimized Random Forest. Applied Sciences. (2018), 8, pp 1859.

[2] Vakharia V, Gupta V K and Kankar P K, Efficient fault diagnosis of ball bearing using ReliefF and Random Forest classifier. *Journal Braz. Soc. Mech. Sci. Eng.* (2017), **39**, pp 2969.

[3] Vakharia V, Kiran M B, Dave N J and Kagathara U, Feature extraction and classification of machined component texture images using wavelet and artificial intelligence techniques, *8th International Conference on Mechanical and Aerospace Engineering (ICMAE)* (2017), pp140-4.

[4] Jin S and Lee S K, Bearing fault detection utilizing group delay and the Hilbert-Huang transform, *Journal of Mech. Sci. and Tech.* (2017), **31**, pp1089–96.

[5] Wu S D, Wu C W, Wu T Y, Wang C C, Multi-Scale analysis based ball bearing defect diagnostics using Mahalanobis Distance and Support Vector Machine, *Entropy*. (2013), **15(2)**, pp 416-33.

[6] Vakharia V and Gujar R, Prediction of compressive strength and portland cement composition using cross-validation and feature ranking techniques, *Construction and Building Materials*.(2019), **225**,pp 292-301.

[7] Emmert-Streib F and Dehmer M., High-Dimensional LASSO-Based Computational Regression Models: Regularization, Shrinkage, and Selection, *Mach. Learn. Knowl. Extr.* (2019), **1**, pp359-83.

[8] Boudiaf A, Bouhouche S, Moussaoui A K and Samira T, An effective method for bearing faults diagnosis, *3rd International Conference on Control Engineering and Information Technology (CEIT) IEEE* (2015), pp1-6.

[9] Loparo K A (2016), Bearing vibration data set. Case Western Reserve University. Available at:www.eecs.cwru.edu/laboratory/bearing.

[10] Tibshirani R, Regression Shrinkage and Selection via the Lasso, *Journal of the Royal Statistical Society: Series B (Methodological)* (1996), **58**, pp267-88.

[11] Hara S. and Maehara, T.. Enumerate lasso solutions for feature selection. In *Proceedings of the 31st AAAI Conference on Artificial Intelligence*. (2017), pp1985–91.

[12] Gujar R and Vakharia V, Prediction and validation of alternative fillers used in micro surfacing mix-design using machine learning techniques, *Construction and Building Materials* (2019), **207**, pp519-27.

IOP Conf. Series: Materials Science and Engineering **841** (2020) 012007 doi:10.1088/1757-899X/841/1/012007

Utilizing Wind Vibrations from Vehicular Airdrag Using Microwind Harnessing Device

Charlyn May A. Gapuz[1] and Ronald Vincent M. Santiago[1]

[1] Mapua University, 658 Muralla St. Intramuros, Manila, Philippines

Abstract. Windbelt has been developed to be an alternative for wind turbines to utilize wind energy, meanwhile, fast-moving vehicles in highways creates wind forces that has potential to be a source of wind energy. This study presents a windbelt prototype and evaluates the output and performance of 1, 5 and 10 connected windbelts when placed on the highway. The results reveal that for every increase in windspeed there is 0.155v and 0.485v increase in output voltage response for 5 and 10 connected windbelt respectively, indicating that the 10 connected windbelt have the highest output voltage response. This indicates that the higher the number of connected windbelts result to larger output voltage.

1. Introduction

Electricity generation in fossil fuel based power plants continuously expands as the demand for power increases. This implies non-stop carbon dioxide emissions which intensified the damage of the ozone layer that led to climate change.

Renewable energy utilization is seen as the solution for this phenomenon. Advantage of this energy source is its availability anytime, anywhere and any season, proper management of this energy source could facilitate the replacement of fossil fuel [1, 2].

Wind turbines are commonly used technology in wind energy production. But it has disadvantages in terms of construction, maintenance and location. Construction of wind turbine will affect land usage and realign the area for energy production rather than crop production [3]. Wind turbines are also sensitive to wind turbulence and gust, other problems usually occur due to the use of gears and bearings [4, 5]. Wind turbines are usually built on coastal or hilly areas and utilizes speed of more than 4m/s for small types, furthermore, the harvesting technologies and equipment for production of renewable energy is costly and location specific, which implies the need for renewable energy generation with simpler, cheaper and less complexity in construction.

With the desire for simple and cheap energy generation another wind energy harnessing device was developed. Windbelt is a device that converts the vibration created by the wind into electrical energy. The main parts of a windbelt device consist of a frame, a membrane and the electromagnetic transducer [6]. The operation starts when a wind causes the flutter/belt to vibrate creating an up and down movement of the magnet between the coils and produces an electrical output (refer to figure 1). It gives many advantages since it does not use gears or bearings like in wind turbines, which are subjected to frictional problems, windbelt can work at either low or high windspeed as it produced 0.6 volts at a minimum windspeed of 2.5 m/s [6].

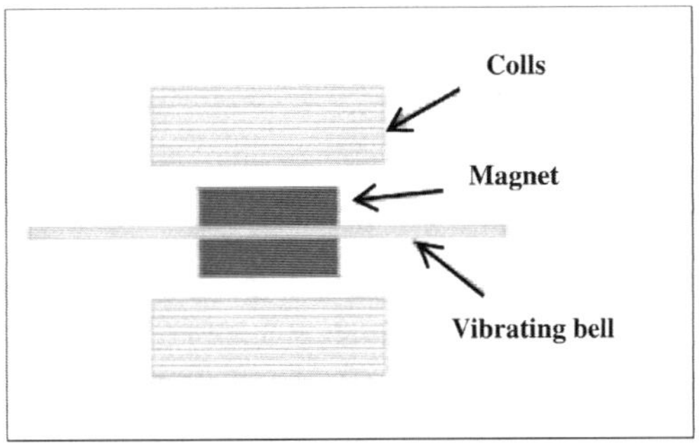

Fig. 1. Windbelt operation.

The concept of energy conversion of windbelt came from Faraday's law of electromagnetic induction which states that "electromotive force (emf) is produced when there is a change in magnetic field caused by the moving conductor". Electromagnetic generator is most utilized in low vibration harvesting according to [7, 8]. Wind belt output performance depends on the type and size of electromagnetic generator [9]. Studies revealed that the flutter/belt of the windbelt produces more output when working in parallel with the wind [10, 11]. A comparison between a single windbelt generator and a wind panel with 5 windbelts was conducted at a the same windspeed of less than 8m/s, the said generators were able to produced 3-5mW and 30-100mW respectively [12].

While there is a naturally occurring wind, there is also a type of wind that is caused by daily human activity, the vehicular airdrag. Vehicular air drag is the strong wind created from the fast-moving vehicles passing along the highway. This generated wind could be a viable source for wind energy generation. The aerodynamic forces created by the vehicle acts on a fixed object on the road, the strong wind forces usually came from big vehicles, hence; vehicle velocity increases near downstream surfaces resulting to higher airdrag which is also dependent on the environmental turbulent levels [13, 14]. These highways have potential as wind energy source that can address the need for electrical energy sources [15].

This study aimed to construct a windbelt prototype that could be used for harnessing electrical energy along the highway, determine the relationship of the number of windbelt prototype and the production of output voltage and determine the differences in output voltage response of 1, 5, and 10 connected windbelt prototypes on straight and slope highways located in the province of Palawan.

2. Methodology

The study was conducted from January-February 2019, in the province of Palawan, Philippines. Locations were selected based from the results of the initial test done between slopes and straight part of the highways. There were few considerations taken during construction of the device since it was intended to be used outdoors; (1) it should not distract the drivers that will cause them to slow down while passing; (2) it should be on a height/appearance that is noticeable to the drivers to avoid accidents; and lastly, (3) the base of the device should be made heavy to resist strong vehicular airdrag.

The fabricated windbelt device, as shown in figure 2, consist of four major parts: 1) Electromagnetic transducer, produces the output from the vibrations created by the belt; 2)Vibrating belt that captures the wind and gets moved when a vehicle passes and creates an airdrag; 3) Frame that serves as the body of the device, where the flutter/belt as well as the electromagnetic transducer is attached; and 4) Base that supports the device so that it may able to stand and resist strong vehicular airdrag, it was made to be detachable for easy and quick set up on the highway.

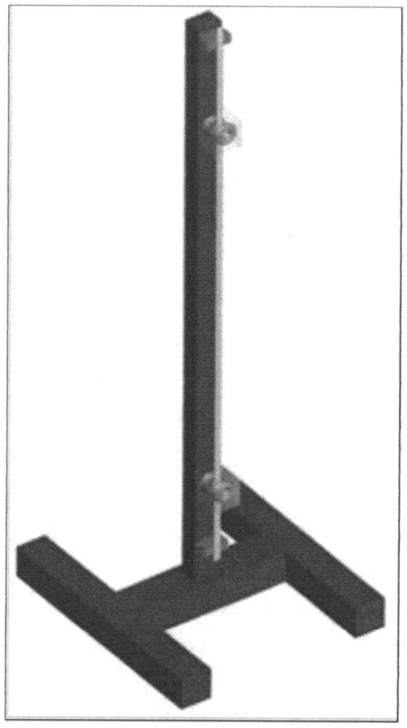

Fig. 2. Fabricated windbelt.

The concept of energy production is shown in figure 3. As the vehicle passes along the road it produces strong vehicular airdrag. The turbulent wind that is created due to fast-moving vehicles along the road was captured by the windbelt, inducing the vibration of the flutter which is in turn converted into electrical energy.

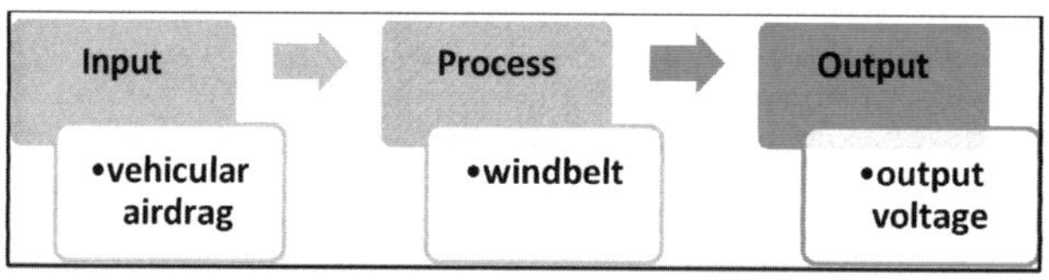

Fig. 3. Conceptual framework.

2.1. *Assembly*
Assembly of the device begins with the construction of the frame made from wood. Then the flutter was then attached and tensioned 60 lbs at both ends of the frame with the use of bolts and nuts, the tension of the flutter was measured using Krikit gauge. Two electromagnetic transducers were created and placed at the near ends of the flutter, each transducer has two loop magnetic coils consisting of 500 turns.

Four magnets were used in each transducer to help increase the change in magnetic flux with the magnetic coils. The magnets were attached to both sides of the flutter and was near the center of the looped coils. The electromagnetic generators were connected in series with the use of wires and alligator clips.

2.2. *Data gathering*
The converted output voltage from the windbelt was measured using potable oscilloscope (JHJDS3012A). The windspeed was measured using cup type anemometer, and all the considered parameters were captured using video camera, manually reviewed and recorded (refer to figure 4).

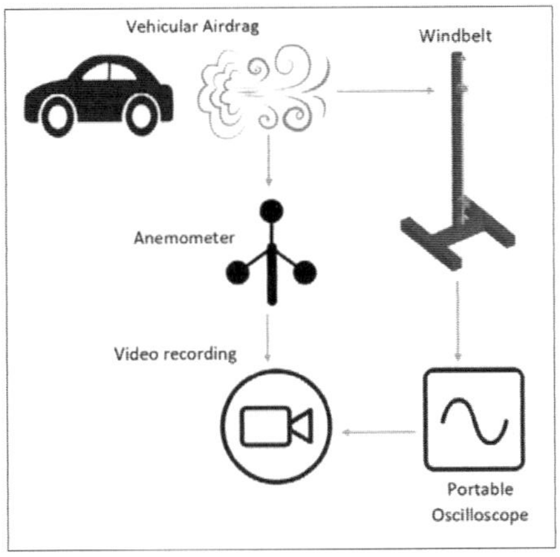

Fig. 4. Data gathering diagram.

3. Results

1, 5 and 10 connected windbelts were tested in 6 highway locations, consisting of 3 straight and 3 slope roads. The test was done in 3 different time periods (7:00-8:00 AM, 12:00-1:00 PM and 5:00-6:00 PM) when most of the vehicles are found on the road. There were 6,732 total number of observations in which 6,729 were interpreted using Analysis of Variance (ANOVA) and regression analysis. The results in table 1 show that the number of connected windbelts contribute to the factors that influences the output voltage.

Table 1. Relationship of the windbelts.

Variable	Df	Sum Sq	Mean Sq	F value	Pr(>F)
WB	2	227.8	113.90	415.7	<2e-16 ***
Residuals	6729	1843.8	0.27		

The comparisons of 1, 5 and 10 windbelt was done using Least Significant Difference (LSD) method and was shown in table 2. The table indicates that the 5 connected windbelt have 0.1935 difference in output voltage with 1 windbelt, while the 10 connected windbelt have 0.4426 difference in output voltage with 1 windbelt and the 10 connected windbelt have 0.2490 difference in output voltage with 5 connected windbelt.

Table 2. Comparison of the windbelt group.

Comparison	Difference	lower.ci	upper.ci	P val
5-1	0.1935	0.1622	0.2248	<2e-16 ***
10-1	0.4426	0.4123	0.4728	<2e-16 ***
10-5	0.2490	0.2185	0.2795	<2e-16 ***

The results of regression analysis revealed that for every increase in windspeed there is 0.485V increase in output voltage response for 10 connected windbelt, which is almost 3 % higher as compared to 0.155V increase of output voltage response for 5 connected windbelt. This indicates that the output voltage increases as the number of connected windbelt increases. The results also indicate that for every unit increase in windspeed the output voltage increases by 0.275V, keeping all the variables constant.

4. Conclusion

The number of windbelt plays an important role in production of electrical energy on highways, connecting more windbelt will affect the production of output voltage. The higher the number of connected windbelts results to an increase of the output voltage. There is an equivalent increase of 0.275V in output voltage every time the windspeed increases. Problems occur with the measurement and recording of other parameters such as output current and power since the measuring device used was not capable of recording/storing data. Good data logger is advisible for more accurate data collection. Boosting the output using a converter and stabilizing it for storage could be further explored.

References

[1] Panwar N, Kaushik S, and Kothari S 2011 Role of renewable energy sources in environmental protection: A review *Renewable and Sustainable Energy Reviews* vol 15 pp 1513-1524 " doi:10.1016/j.rser.2010.11.037"

[2] Leung D Y and Yang Y 2011 Wind energy development and its environmental impact: A review *Renewable and sustainable energy review* vol 16 pp 1031-1039 "doi:10.1016/j.rser.2011.09.024"

[3] Jaber.S 2013 Environmental impacts of wind energy *Journal of Clean Energy Technologies* vol 1 pp 251-254 "DOI:10.7763/JOCET.2013.V1.57"

[4] Chiras D 2010 *Wind power basics: a green energy guide* Gabriola Island, Canada pp 24-29

[5] Tong W 2010 *Wind power generation and wind turbine design* Southampton, Boston, pp 28-40

[6] Frayne S M 2009 Generator utilizing fluid-induced oscillations US Patent US20090309362A1

[7] Beeby S P, Torah R, Tudor M, Glynne-Jones P, O'donnell P, Saha C and Roy S 2007 A micro electromagnetic generator for vibration energy harvesting *Journal of Micromechanics and microengineering* vol 17 pp 1257 "doi:10.1088/0960-1317/17/7/007"

[8] Cepnik C, Lausecker R and Wallrabe U 2013 Review on electrodynamic energy harvesters—a classification approach *Micromachines* vol 4 pp 168-196 "doi:10.3390/mi4020168"

[9] Zhu D, Beeby S P, Tudor M J, White N M and Harris N R 2013 Novel miniature airflow energy harvester for wireless sensing applications in building *IEEE sensors journal* vol 13 pp 691-700 "DOI:10.1109/JSEN.2012.2226518"

[10] Li S and Lipson H 2009 Vertical-stalk flapping-leaf generator for wind energy harvesting *ASME 2009 Conf. on Smart Materials, Adaptive Structures and Intelligent Systems* Oxnard, California, USA pp 611-619 "doi:10.1115/SMASIS2009-1276"

[11] Zhu D 2011 Vibration energy harvesting: machinery vibration, human movement and flow induced vibration *Sustainable Energy Harvesting Technologies-Past, Present and Future* Dr Tan Y K Ed IntechOpen chapter 2 pp 25-54

[12] Quy V D, Van Sy N, Hung D T and Huy V Q 2016 Wind tunnel and initial field tests of a micro generator powered by fluid-induced flutter *Energy for Sustainable Development* vol 33 pp 75-83 "doi:10.1016/j.esd.2016.04.003"

[13] Duncan B, D'Alessio L, Gargoloff J and Alajbegovic A 2017 Vehicle aerodynamics impact of on-road turbulence *Proc. of the Institution of Mechanical Engineers, Part D: Journal of Automobile Engineering* vol 231 pp 1148-1159 "DOI: 10.1177/0954407017699710"

[14] Lichtneger P and Ruck B 2013 Transient wind loads on flat elements induced by passing vehicles *Fachtagung "Lasermethoden in der Schwingungstechnik" Proc. of Technical Field Meeting "Laser methods in vibration engineering* Munich, Germany pp 1-12

[15] Atallah R, Khabbaz M and Assi C 2016 Energy harvesting in vehicular networks: A contemporary survey. *IEEE Wireless Communications* vol 23 pp 70-77 "DOI: 10.1109/MWC.2016.7462487"

Acknowledgment

The authors gratefully acknowledge and thank the DOST-ERDT for the financial support to this study.

IOP Conf. Series: Materials Science and Engineering **841** (2020) 012008 doi:10.1088/1757-899X/841/1/012008

Creation of a Driverless Electric Cargo Vehicle with a Modern Energy Storage System

K Karpukhin[1] **and A Terenchenko**[1]

[1] NAMI Russian State Scientific Research Center, 2 Automotornaya st, Moscow, Russia, 125438

E-mail: K.Karpukhin@nami.ru

Abstract. The paper describes development of an experimental cargo electric vehicle equipped with automated driving features. The focus is made on the powertrain design. In particular, a description of the electrical system is given detailing its modular design. The latter allowed for convenient arrangement of the units and devices intended for energy storage and feeding of high voltage and low voltage components of the traction system and auxiliary systems of the vehicle. The bulk of the paper describes the key features of the design process beginning with CAD models, from which the vehicle chassis and the powertrain were designed, proceeding to selection of the powertrain components, and concludes with the description of the vehicle produced in accordance with the elaborated design. The paper also gives a survey of automatic driving functions being implemented within the vehicle at the present stage of development, including path tracking, lane keeping, keeping safe distances to surrounding objects, and some others.

1. Introduction

Personal mobility is a basic need of humans. In 1998, we had a total about 700 million vehicles, but in 2019, we have over 1 billion vehicles in the world. Hybrid and electric vehicles developed differently in the last 30 years, but they had hard way into the market. According to forecasts of the European Commission for transport by 2020 in Europe 7 % of all new passenger and light commercial vehicles will be electric, increasing to a massive 31% by 2030. This measure is a forced connection with the sharp deterioration of the situation on the planet associated with air pollution and global warming. The solution of these problems is the creation of electric vehicles, production of which is established in the leading countries of the world [1, 2, 3].

Nowadays, the share of electric vehicles in the global fleet constitutes 1-2%. Great Britain and France have declared the ban for the gasoline and diesel vehicle sales would be introduced by 2040. The similar announcement was made by the Ministry of Industry of China, which is conducting researches of this issue. In the same time, China has intention to become the world leader in production of electric vehicles.

The main obstacle preventing electric vehicles from large scale commissioning is the cost issue. Prices of “low-budget” passenger EVs start from $30 000, not to say about commercial EVs, which are far more expensive. Meanwhile their conventional counterparts are at least twofold cheaper.

Expert estimates suggest that electric vehicles will be able to compete with conventional ones when the kilowatt per hour price of lithium ion batteries drops from current $250 (and above) to $100. In the expert opinion, this should happen by 2026. Note that back in 2010 the kWh price was $1000. Other complications also exist. The charging infrastructure is far from mature; battery price drop may be slower than expected; the governmental policy may change. For example, the Chinese government imposed a ban for foreign investments in the production of lithium ion accumulators and the most valuable components thereof, i.e. cathodes.

The Russian EV market and fleet are still small – statistical data states that by 1 March 2019 only 3600 electric vehicles were in commission. Unlike China, EU and USA, the Russian ecology policy doesn't provide sufficient stimulating measures in respect of EV. Buyers and owners of electric vehicles aren't granted with decent privileges. As a result, the demand for electric vehicles is small entailing low production volumes as well as high prices thereof. Another issue complicating the expansion of EV in Russia is specifics of regional climate. In some regions, air temperature may vary from -40 to 40°C throughout the year [4, 5, 6].

Despite these complications, Russian vehicle manufacturers conduct R&D projects in order to advance EV technology into the market. One of such projects is conducted by NAMI Russian State Scientific Research Center in collaboration with a vehicle manufacturer "KAMAZ". In this project, a heavy-duty electric vehicle is being developed having state of the art electric storage system and also equipped with automated driving features. The sequel of this paper describes the powertrain of this vehicle and gives a survey of automated driving features being implemented within the vehicle at the current stage of the project [7, 8].

2. Powertrain design of the experimental electric vehicle

The traction electric drive was integrated into the mechanical transmission and mounted on the chassis "KamAZ-4308". Figure 1 shows a model of the chassis including the powertrain, which consists of the electrical system (1), and the traction electric motor (2) connected to the driving axle (4) by a cardan shaft. The axle 5 is of driven type. All the components are mounted on the chassis frame 3.

Figure 1. The chassis of the experimental electric vehicle.

The said electrical system is shown in figure 2. It includes the traction battery (1) organized in modules, the battery charger (2) with the plug socket (3), high voltage (4) and low voltage (5) power electronics. It also includes a number of voltage converters for feeding the auxiliary systems. In particular, DC/AC converter 6 feeds the motor driving the air compressor. DC/AC converter 7 provides energy supply to the system responsible for thermal regulation of the powertrain components. One of the DC/DC converters 8 feeds the electric motor driving the pump of the power steering system. Other converters 8 provide power supply for the low voltage components of the thermal management system, low voltage part of the traction electric equipment, and other auxiliary loads of the vehicle. All the components of the electric system are installed within the common frame. The latter and the modular design of the electrical system allow assembling of this system independently from the vehicle chassis. In turn, this allows for minimal complexity of the assembling process and technological tools employed, thus entailing reduced assembling time.

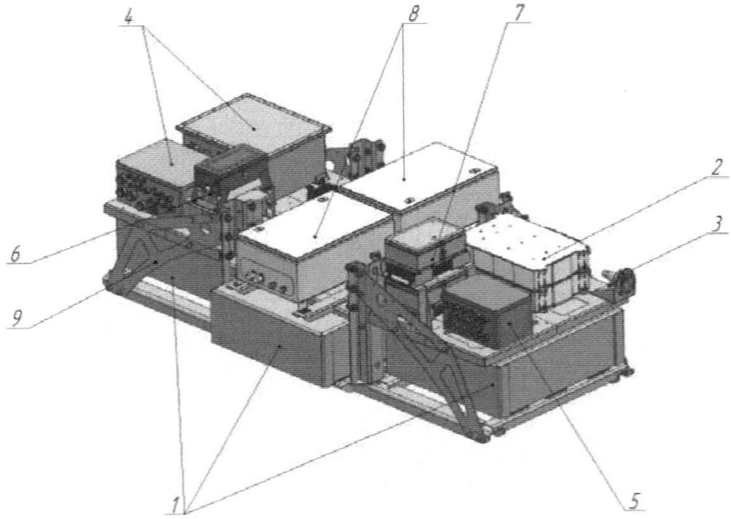

Figure 2. Design of the modular electric system.

The used traction electric motor TM4 SUMO HD (figure 3) is a permanent magnet electric machine with 195 kW continuous power and 2060 Nm continuous torque. The motor is mounted on the chassis frame by means of additional cross-members. The inverter is placed between the traction motor and the battery on the specially designed cross-members.

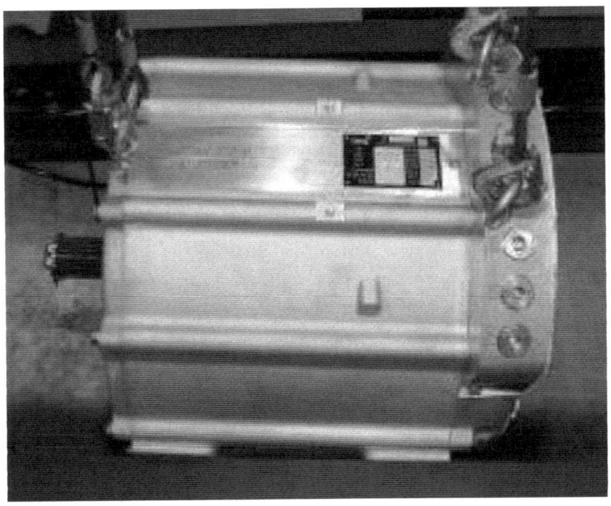

Figure 3. Traction electric motor (TM4 SUMO HD).

The traction battery consists of three modules, whose characteristics are listed in Table 1.

The traction battery modules are placed in the bottom of the frame. This allows for lowering the vehicle center of gravity, which is relevant for reducing risk of rollovers. In order to prevent harmful impact of the environment and road unevenness, the battery modules are installed on a metal platform providing protection thereof [9].

The charging device (EVO22KL) implements an interface of the traction battery with the standard electric grid 220V/50Hz. The charger location provides an easy access to the device for performing maintenance operations. The casing of the charging device ensures protection level complying with IP56 rating.

Design of the described vehicle implies no cab since it is supposed to operate either in automated or remote-controlled mode. The vehicle was constructed and equipped with the described electric powertrain. Figure 4 shows a general view of the vehicle.

Table1. Specification of the traction battery and its modules.

Manufacturer	Microvast
Model, number of modules	MV-B, 2 MV-C, 1
Battery voltage, V	437-680.5
Energy content, kWh	75.5
Dimensions of the module, mm (WxHxD)	1060x240x660 820x240x660
Life span of the battery (20% capacity drop)	300 000 km, 8 years

Figure 4. The developed autonomous electric vehicle.

3. Automated driving functions implemented within the experimental vehicle

The following automated driving functions are being implemented within the described vehicle [10]:

- tracking of a predetermined path;
- tracking of specified speed and keeping safe distances between the vehicle and surrounding objects;
- automatic braking preventing collisions with obstacles, automatic emergency braking;
- electronic stability control and rollover prevention;
- recognition of traffic signs and lane markings;

The path-tracking feature provides an automatic control of vehicle directional motion based on data received from the machine vision tools (cameras and radars) and/or from the navigation system. The way of using these data sources depends on the visibility of lane markings. Allowed deviation of the vehicle from the lane center is 0.5 m. The operating velocity range of the feature is 1-25 km/h.

Velocity tracking, distance keeping, and automatic braking functions operate within the same velocity range as the path tracking. The factors taken in account by these functions are speed limits established within the operational domain of the vehicle and obstacles detected in the heading direction of the vehicle. The automatic emergency braking feature activates when the vehicle control system detects a fault considered critical in respect of driving safety. In that case, vehicle decelerates at a moderate constant rate (not exceeding 2 m/s^2) until the full stop, turning on the hazard flashers.

The stability control feature prevents excessive understeering and oversteering of the vehicle while maneuvering (e.g. negotiating a turn). In order to do this the feature employs the regenerative braking system regulating the vehicle speed and yawing moment exerted by the driving wheel tire forces. The feature also prevents the driving wheels from excessive slip, thus providing sufficient reserve of the lateral tire force during directional maneuvering.

4. Conclusions and future work

The described electric vehicle has been constructed in accordance with the elaborated design of the chassis and powertrain. A distinctive feature of the developed powertrain is the electrical system with a modular structure. This system provides an energy storage for both traction needs and feeding of the onboard auxiliary systems. Interfaces between the components with different voltage levels and forms were implemented by means of DC/DC and DC/AC converters included in the said electric system. At the current stage of the project, the automated driving functions are being implemented within the vehicle controller. The next stage implies testing of the vehicle with respect to functioning of the powertrain and the automatic driving features.

5. References

[1] Miller M J 2010 *Propulsion systems for hybrid vehicles 2nd Edition.* UK: The Institution of Engineering and Technology.

[2] International Energy Agency. Hybrid and Electric Vehicles *M. The Electric Drive Chauffeurs*, 2017.

[3] 2014 *Encyclopedia of Automotive Engineering M.* John Wiley & Sons ISBN: 978-0-470-97402-5 Online.

[4] Terenchenko A, Karpukhin K, Kurmaev R 2015 *Features of operation of electromobile transport in the conditions of Russia* J. Paper of EVS 28 International Electric Vehicle Symposium and Exibition, KINTEX, Korea.

[5] Gaines L, Cuenca R, Life-Cycle Costs of Lithium-Ion Vehicle Batteries 2000 *J. SAE Technical papers* **2000-01-1483**.

[6] AL-Refai A, Rawashdeh O, Abousleiman R 2016 An Experimental Survey of Li-Ion Battery Charging Methods *J. SAE International Journal of Alternative Powertrains*, **5 (1)**.

[7] Kulikov I.A., Shorin A.A., Bakhmutov S.V., Terenchenko A.S., Karpukhin K.E. 2016 A Method of Powertrain's Components Sizing for a Range Extended Electric Vehicle. *J. SAE Technical Papers* **2016-01-8096**.

[8] Singh S, Mathur A, Das S, Sinha P, et al 2017 Development of Smart Public Transport System by Converting the Existing Conventional Vehicles to EV's in Indian Smart Cities. *J. SAE Technical Papers* **2017-01-2011**.

[9] Shorin A A, Karpukhin K E, Terenchenko A S, Kondrashov V N 2018 Traction module of cabless unmanned cargo vehicles with electric drive *J. International Journal of Mechanical Engineering and Technology* **9 (11)** 1903–1909.

[10] Ramesh K., Siddhesh N., Kundan S., Deepak V. 2013 Data Acquisition System And Telemetry System For Unmanned Aerial Vehicles For Sae Aero Design Series. *J. International Journal of Electronics and Communication Engineering & Technology* **4 (5)**, pp. 90-100.

Acknowledgments

The paper was prepared under the agreement # 14.624.21.0049 with the Ministry of Education and Science of the Russian Federation (unique project identifier RFMEFI62417X0049).

IOP Conf. Series: Materials Science and Engineering **841** (2020) 012009 doi:10.1088/1757-899X/841/1/012009

Definition of Contact Stresses and Gaps in the Hinges of Chain Gears

Yu V Belousov[1], **S L Shambina**[2] **and F V Rekach**[2]

[1]Bauman Moscow State Technical University, 5, Baumanskaya 2-ya Street, Moscow, 105005, Russian Federation
[2]Peoples' Friendship University of Russia (RUDN University), 6 Miklukho-Maklaya Street, Moscow, 117198, Russian Federation

Email: belou.80@mail.ru

Abstract. The article discusses the issues of determining contact stresses in the hinges of chain gears. It is necessary to know these stresses to assess the hinges' wear resistance, since the wear of hinges (joints) is the most common cause of chain gears' damage. In this regard, the focus was made on determining the necessary gap between the bushing and the roller of the chain gears. This gap significantly affects the largely determines the distribution pattern and magnitude of the stresses in the hinge, and, consequently, the wear resistance of chain gears. A method for calculating the maximum contact stress in the hinges of chain gears has been developed. It was proved that a decrease of the gap in the hinges influence in the following way: the contact angle between the roller and the bushing increases and the maximum contact stresses become lower. However, for real contact angles, the contact stresses greatly exceed the permissible average specific pressure, causing increased wear on the joints (hinges). One of the possible ways to solve this problem is to use transitional fittings in hinges, in which the gaps and tightness are relatively small. When pressing the rollers into the bushings, microroughnesses are cut from the contact surfaces, and the tightness in the connection practically disappears. Some examples of determining gaps and contact stresses in the hinges of chain gears are given.

1. Introduction

Wear of hinges is the most common type of damage of chains in closed and semi-closed gears of machines, engines and general engineering equipment. Therefore, the main calculation of chain gears is carried out on the condition of wear resistance of the chain joints. For the calculation, it is necessary first of all to know the contact stresses between bushings and rollers of the chain. However, the determination of stresses in the contact zones is a rather complicated task. Therefore, the calculation is carried out according to the conditional pressure in the chain hinges under the assumption of a zero gap between the roller and the bushing and a uniform distribution of pressure in the hinge. Moreover, there is no data on the size of the required gap in the chain hinges in State Standard of Russia (GOST) 13568-97 for driving roller chains and in GOST 588-81 for the traction plate chains, as well as in the literature on this issue. This gap mainly determines the pattern of distribution and meanings of contact stresses in the hinge, and thus wear resistance of chain gears.

2. Calculation of contact stresses

For hinges of chain the bore diameter of the bushing is: $d_{bore}=d+S_h$,

where d is a roller diameter, S_h is a gap between a bushing and a roller in a hinge.

Gaps between the bushing and the roller in the chain hinges are very small. In this regard, the calculation can be reduced to the problem of indenting a circular finger into a body with a circular cavity, which, after application of the load, touches the finger over a relatively large area. The conjugation parameters here can be determined on the basis of the solution of the contact task of the theory of elasticity on the internal compression of two cylindrical bodies whose radii are almost equal. The solution to the task of compressing of cylindrical bodies with similar radii taking into account friction can be approximated by the same task without friction, since the friction forces on the contact surface lead to a redistribution of the contact pressure diagram and its displacement in the direction opposite to the rotation of the gear roller. In this case, the displacement angle ρ of the center of the contact arc is equal with high accuracy to *arctg(f)*, and half of the contact angle φ_0 quite not depends on the friction coefficient *f* [1, 2, 3, 4].

To determine the stresses in the gear hinges when $d_{bush}/d>1,5$, where d_{bush} is a bushing's diameter, we take the following pressure distribution on the contact surface:

$$p(\varphi)=C\cdot\cos\left(\frac{\pi\varphi}{2\varphi_0}\right), \tag{1}$$

where φ is an angular current coordinate. Whereas maximum contact stress at the hinge (when $\varphi=0$) $\sigma_{max}=C$. The value *C* can be found from the roller equilibrium conditions:

$$\int_{-\varphi_0}^{\varphi_0} p(\varphi)\cos\varphi d\varphi=\frac{2N_{sp}}{d}, \tag{2}$$

where $N_{sp}=N/l$ is a specific contact load (N is the total load on a hinge, l is bushing's length). That is

$$\int_{-\varphi_0}^{\varphi_0} C\cdot\cos\left(\frac{\pi\varphi}{2\varphi_0}\right)\cos\varphi\, d\varphi=\frac{2N_{sp}}{d}. \tag{3}$$

Whence

$$C\frac{\pi\cos\varphi_0\varphi_0}{0,25\pi^2-\varphi_0^2}=\frac{2N_{sp}}{d}. \tag{4}$$

When $\varphi_0=0,5\pi$, that is when the diameters of the roller and the bore of the bushing are equal, the "uncertainty" of the type 0/0 arises on the left side of expression (4). To disclose the uncertainty, we use the Lopital rule:

$$\lim_{\varphi_0\to 0,5\pi}\left(\frac{\pi\cos\varphi_0\varphi_0}{0,25\pi^2-\varphi_0^2}\right)=\lim_{\varphi_0\to 0,5\pi}\left[\frac{\left(\pi\cos\varphi_0\varphi_0\right)'}{\left(0,25\pi^2-\varphi_0^2\right)'}\right]=\frac{\pi}{2}. \tag{5}$$

In this case if $\varphi_0=0,5\pi$:

$$\sigma_{max}=C=\frac{4}{\pi}\frac{N_{sp}}{d}=\frac{4}{\pi}p, \tag{6}$$

where $p=N_{sp}/d$ is the average specific contact pressure in the hinge.

Expression (6) contains a multiplier ($4/\pi$) that takes into account the sickle-shaped distribution of pressure around the circumference of the pressure, the same as when calculating connections with tightness, which perceive the bending moment. From expression (4) we get:

$$C=\frac{2N_{sp}}{d}\frac{0{,}25\pi^2-\varphi_0^2}{\pi\cos\varphi_0\varphi_0}. \tag{7}$$

Now we express the maximum contact stress by the average specific pressure in the hinge:

$$\sigma_{max}=C=kp\ ;\quad k=2\,\frac{0{,}25\pi^2-\varphi_0^2}{\pi\cos\varphi_0\varphi_0}. \tag{8}$$

The calculation of the maximum contact stress in hinges of a chain gear which was made using formula (8) agrees quite well with the calculation according to the existing methodology for sliding bearings. For comparison, we calculate the coefficient *k* using the formula (8) and compare it with a similar coefficient for sliding bearings given in [5]. The calculation results are shown in Fig. 1.

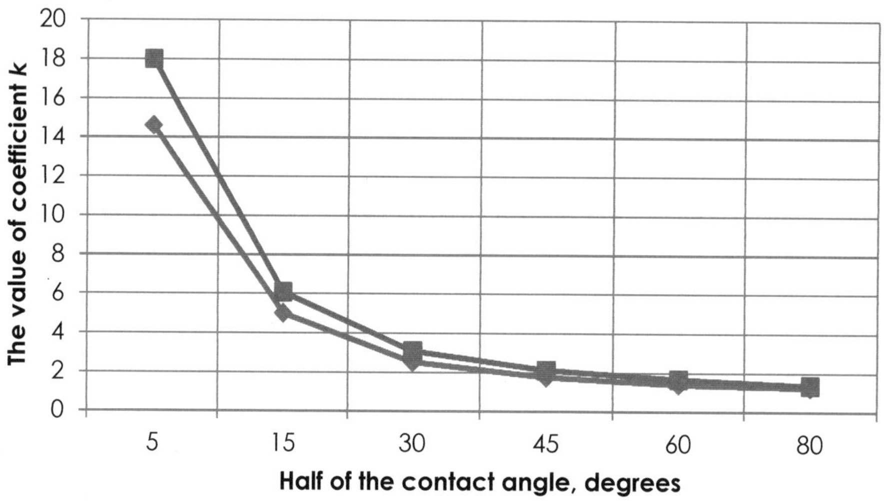

Figure 1. The dependence of coefficient k on half of the contact angle φ_0.

Here, the upper line corresponds to the calculation according to formula (8), and the lower one, according to the data of [5]. From Fig. 1 it can be seen that the value of the coefficient k determined by formula (8) is slightly higher, especially for small contact half-angles φ_0. For large half-angles φ_0 the calculations practically coincide. The maximum discrepancy does not exceed 20%. However, the dependences for calculating given in [5] are very complex, and their results are presented in a table form. And the resulting formula is simple and convenient for calculation.

3. Determination of gaps in hinges

For calculating the contact stress it is necessary to know the value of the angle φ_0. The angle φ_0 is a function of the loading coefficient β. In this case (when both the roller and the bushing are made of steel) it can be determined by the following formula:

$$\beta=\frac{2N_{sp}}{\pi\varepsilon_{av}}\frac{1-\nu^2}{E}, \tag{9}$$

where ε_{av} is the average radial gap in the hinge; E and ν are modulus of elasticity and Poisson's ratio of steel, respectively. Then the average relative gap clearance ψ_{av} in the hinge is equal:

$$\psi_{av}=\frac{4p}{\pi\beta}\frac{1-\nu^2}{E}, \qquad \psi_{av}=\frac{S_{av}}{d}=\frac{\varepsilon_{av}}{r}. \tag{10}$$

For maximum load:

$$\psi_{av}=\frac{4[p]}{\pi\beta}\frac{1-\nu^2}{E}, \tag{11}$$

where $[p]$ is the permissible average specific pressure in the hinge.

The permissible average specific pressure in the hinges of roller chains [6] is:

$$[p]=\frac{[p_0]}{K_{op}}, \tag{12}$$

where K_{op} is an operating ratio. Pressure $[p_0]$ depends on chain pitch and the frequency of small sprocket's rotation. The highest value $[p_0]$ = 35 MPa corresponds to the rotation speed of a small sprocket n<50 min-1. The minimum value for the operating coefficient the lubricated transmission coefficient for lubricated gears K_{op}=0, 8. Then $[p]_{max}$=43, 75 MPa.

As already noted, the angle φ_0 is a function of the loading coefficient β, which is defined graphically in [5]. Using this dependence, the average relative gap ψ_{av} was calculated for $[p]_{max}$=43, 75 MPa and for various contact half-angles φ_0. The calculation results are shown in Fig. 2.

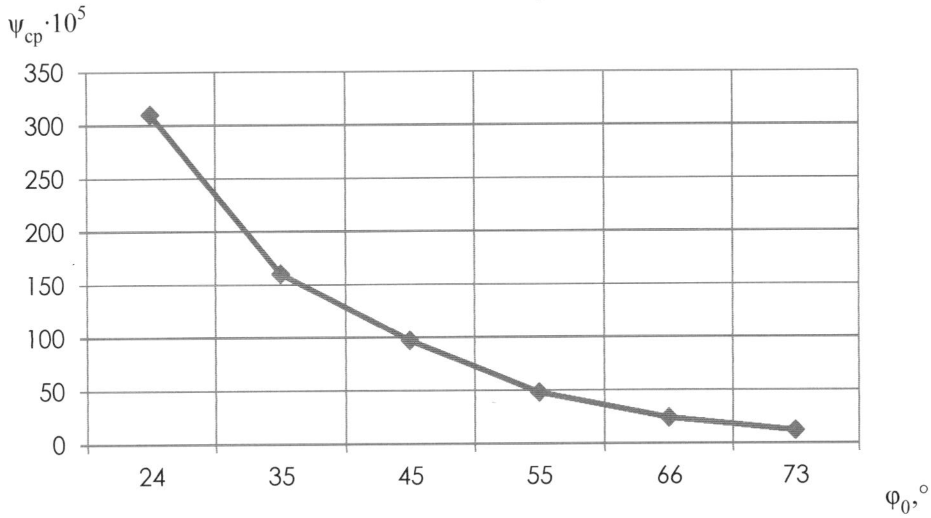

Figure 2. Dependence of the average relative gap ψav on half the contact angle φ_0.

For $\varphi_0>45^\circ$ it is technologically difficult to provide the required gaps in the hinge. If we also take into account that for $p<[p]_{max}$ or even for $p<[p]$, the relative gaps in the hinge decrease. Therefore, we will further assume that $\varphi_0\leq45^\circ$. For this range, we can propose the following dependence connecting the contact half-angle φ_0 with the average relative gap ψ_{av}:

$$\psi_{av}\cdot10^5=\frac{1{,}8\cdot10^6}{\varphi_0^{2{,}78}}+51{,}4\,. \tag{13}$$

To determine the contact half-angle itself φ_0 for $\varphi_0\leq45^\circ$, the following dependence can be proposed:

$$\varphi_0=94\beta^{0{,}53}. \tag{14}$$

In this case, the average diametrical gap is $S_{av}=\psi_{av}d$. For example, we define the contact half-angle φ_0 and contact stress for the case when p=20 MПa and $S_{av}=0{,}006d$.

In this case:

$$\beta=\frac{4\cdot 20}{\pi\cdot 0{,}006}\frac{1-0{,}3^2}{2{,}1\cdot 10^5}=0{,}0184 \quad \text{and}\ \varphi_0=94\cdot(0{,}0184)^{0{,}53}=11{,}31^{\circ}.$$

The maximum contact stress in the hinge is:

$$\sigma_{max}=C=kp=2\,\frac{0{,}25\pi^2-(11{,}31\pi/180)^2}{\pi\cos(11{,}31^{\circ})(11{,}31\pi/180)}\,20=160\ \text{MPa}.$$

It greatly exceeds the permissible average specific pressure, causing increased wear of the joints. This tendency remains the same for the entire range of angles $\varphi_0\leq 45^{\circ}$.

It is possible to reduce the contact stress in the hinges of the chain gear by increasing the half-angle φ_0 of the contact of the roller with the bushing, i.e. increasing the loading coefficient β. This is achieved by reducing the average radial gap in the hinges ε_{av}. In this case, for the range of angles $\varphi_0>45^{\circ}$, we can propose the following dependence for their determination:

$$\varphi_0=63{,}46\beta^{0{,}198}. \qquad (15)$$

4. Conclusions

The necessary gaps in the hinges are very small, and it is quite difficult to provide them technologically. One of the possible ways to approach to solve this problem is to use transitional tightness in hinges, for example, of the type H7/m6 or H7/k6. In these cases, the probability of obtaining a gap and a preload is approximately the same. Gaps and preloads are relatively small. When pressing the rollers into the bushings, microroughnesses are cut from the contact surfaces, and the tightness in the connection practically disappears, since it is here approximately equal to the correction for a hug (cut) of microroughnesses during assembly.

5. References

[1] Belousov Yu V Analysis conditions for reliable attach of workpieces during lathe machining on the mashines with numerical soft-ware control. 2017 *RUDN J. of Eng.Res.* **19** 1 91

[2] Belousov Yu V Modelling of force interaction of the tool during mechanical machining. 2016 *J. Struct. Mech. of Eng. Constr. and Build.* **4** 62

[3] Shambina S L, Rekach F V and Belousov Yu V On new modifications of some strength criteria for anisotropic materials 2016 *J. Key Engineering Materials* **724** 53

[4] Hoffman N P and Stolz V Ontransient growth of wear pattern properties 2010 *J. Wear* **268** 7-8 886

[5] Kogaev V P and Drozdov Yu N 1991 *Prochnost' i I znosostoikost' detalei mashin [Strength and durability of machine parts]. Study guide for mechanical engineering higher education institutions* (Moscow: Vysshaya Shkola Publ.) p 319

[6] Andrienko L A, Baykov B A and Zakharov M.N. etc. 2014 *Detali mashin [Machine parts]. Textbook for universities / Under the editorship of O. A. Ryakhovsky.* (Moscow: Baumanpress Publ) p 465

Acknowledgments

This paper was financially supported by the Ministry of Education and Science of the Russian Federation on the program to improve the competitiveness of Peoples' Friendship University of Russia (RUDN University) among the world's leading research and education centers in the 2016-2020. This publication was prepared with the support of the "RUDN University Program 5-100".

IOP Conf. Series: Materials Science and Engineering **841** (2020) 012010 doi:10.1088/1757-899X/841/1/012010

Study of the Attenuation Force Generated by a Magnetorheological Fluid in Industrial Robot Grippers Shock Absorbers

Juan Carlos Valencia Salinas[1,a], **Diego Alejandro Jumbo Auquilla**[1,b], **Marco Singana**[1,c], **Caterine Donoso**[1,d]

[1] Departamento de Ciencias de la Energía y Mecánica, Universidad de las Fuerzas Armadas ESPE, Av. General. Rumiñahui s/n, Sangolquí, P. O. Box 171-5-231B, Ecuador

E-mail: ajcvalencia5@espe.edu.ec

Abstract. The attenuation force generated by the use of magnetorheological fluid (FMR) during the expansion and compression of the shock absorber in robot grippers has been studied. The shock absorber has a stroke of 25 mm and a volume of 4 071.50 mm^3 in which the FMR made up of ferric micro-particles and nanoparticles is housed; this fluid modifies its rheological properties when exposed to a magnetic field of variable intensity generated by an electromagnet. The intensity of the magnetic field is gradually varied in order to obtain an attenuated and controlled dynamic force in the dissipation of energy. The research includes an analysis of magnetic field generation with the purpose that the particles of the fluid are aligned according to the polarity of the flow lines, thus forming a fibrillary structure chained in a few milliseconds. This magnetic field modifies the viscosity of the fluid, taking it from a low viscosity to a higher density state without altering its chemical composition and shape. The results are reflected in the direct correlation and in the obtaining of data where the saturation of the fluid is given to a maximum intensity of magnetic flux (B) of 0.01 Tesla; and a damping force of 6,12 Newtons. This concludes that the theoretical and experimental analysis of attenuation in FMR based buffers serve to mitigate impact excitations, in order to provide a buffered velocity of motion.

1. Introduction

The application of FMR in the industrial field has been advancing with the passing of the years, today many investigations are carried out as in ascent valves, control pumps, brakes, assembly claws, among others [1]. The study of magnetorheological fluids is based on the ability to change instantly from a viscous fluid state to a semi-solid state in the presence of a magnetic field. FMR fluids have iron microparticles that are held together by the magnetic field forming a kind of chain that is resistant to a certain level of shear stress by behaving like a semi-solid material [2]. The function of a shock absorber starts when it receives a series of oscillations produced by irregularities in the medium, then the energy received by the shock absorber is dissipated when the shear stress exceeds a critical value, being beneficial to design a magnetorheological shock absorber FMR to control the attenuation when operating a gripper coupled to an industrial robotic arm [3].

2. Experimentation

2.1. Power supply
The shock absorber required a source that provides 3 independent outputs with ranges of 0-30 volts and 0-3 amps, a serial output was configured to obtain the value of 12 volts and 0.92 Amps of current [4].

2.2. Magnetic Field Sensor
The measurement of the magnetic field is possible thanks to the magnetic field sensor Xplorer Glx of the brand Pasco, this magnetic sensor offers sensitivity in a single range of -1000 to +1000 gauss with an accuracy of ± 3 gauss [5].

2.3. Dynamometer
The response to attenuation is reflected thanks to the Hoosiwee Weighing Machine, which has a large capacity of electronic scales, has a high precision sensor system that allows you to measure a maximum capacity of 50 kg to ensure a precise weighing session [3].

2.4. Magnetoreological fluid
The LORD MRF-140CG brand magnetorheological fluid is used with a density of $\rho = 3740$ kg m^{-3} and viscosity of μ=0.28 Pa*s [6].

2.5. Methodology
The investigation included a design stage of the shock absorber and winding, then the acquisition of a commercial FMR and finally the application in grippers for the subsequent study of the attenuation force results.

2.6. Shock absorber design
The mechanical design of the shock absorber was oriented towards a short-stroke suspension system where a maximum range value of 0.025 m was established for the piston rod in expansion or compression. The body of the shock absorber was constructed of 6061 T6 aluminum, (see figure 1) [3].

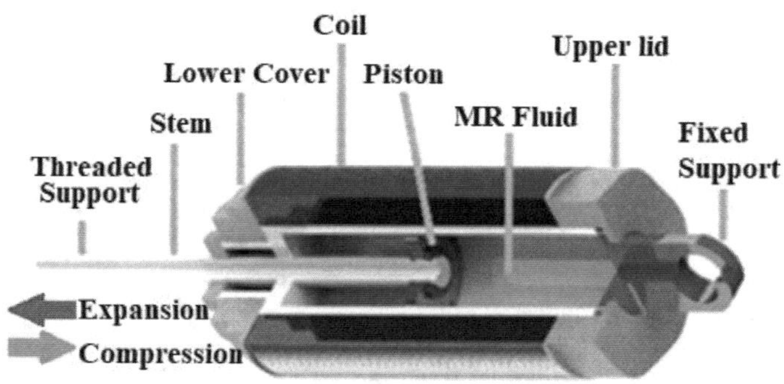

Figure 1. Diagram of the components constituting the shock absorber

The static structural analysis of the shock absorber is done by Computer Aided Engineering (CAE) and Finite Element Analysis (FEA).The maximum deformation of the shock absorber cylinder shown in the figure 2; and piston rod shown in the figure 3 are very low, considered as negligible and acceptable. The resulting stress of Von Mises is relatively low compared to the yield strength of 6061-T6 aluminum, which has a value equal to 315 [MPa], so it is concluded that the design is safe. To consider an acceptance of the design we note that the critical safety factor of the design is greater than the permissible safety factor of $N = 2$, the design being considered as safe.

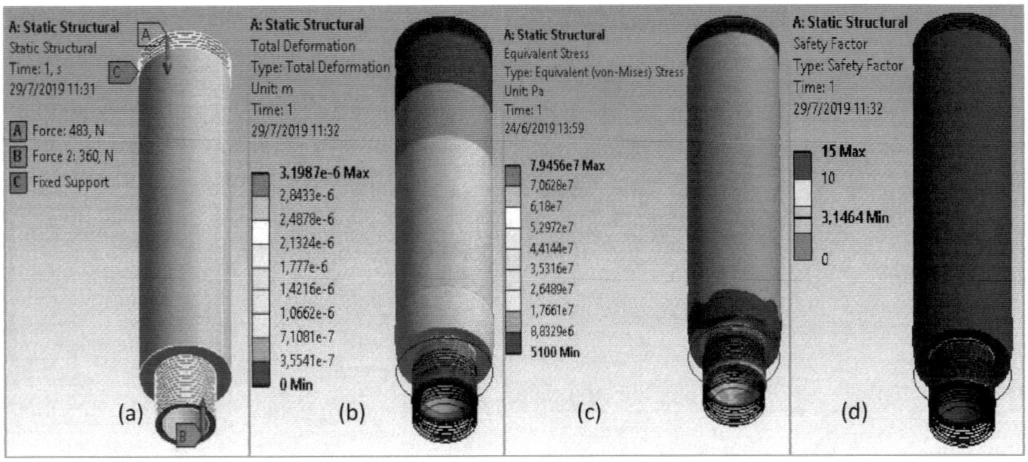

Figure 2. (a) Application of forces and supports on the shock absorber cylinder. (b) Deformation. (c) Von-Mises Efforts. (d) Security factor.

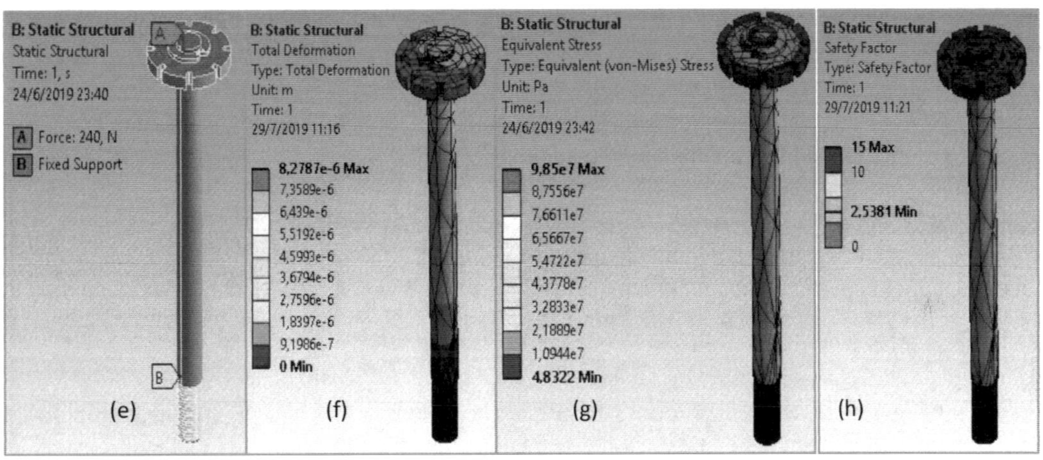

Figure 3. (e) Application of forces and supports on the piston rod of the shock absorber. (f) Deformation. (g) Von-Mises Efforts. (h) Security factor.

2.7. Winding

The technique of indirect-external winding was used as can be seen in figure 4, the length of the coil Ls=0.036 m, with 6 layers and a number of coils N=56, using AWG 22 copper wire of diameter Dal=6.44 $\times 10^{-4}$ m and resistivity 5.6/100 m, as in equation (1) [4].

$$N = \frac{L_S}{D_{al}} \tag{1}$$

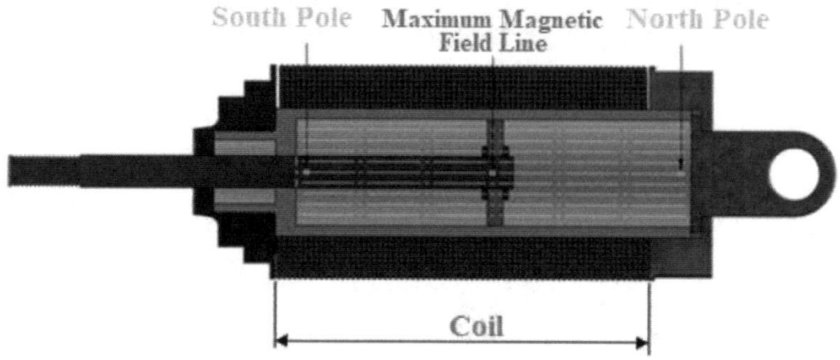

Figure 4. Indirect-External Rewind

2.8. Behaviour of the FMR in the shock absorber

The flow rate (Q) generated by the fluid passing from one chamber to another within the internal cavity of the shock according to equation (2) was determined as a function of the speed of the shock absorber; the transverse area of the Aa shock absorber is obtained from the following equation (3); and the piston diameter Dp of 0.012 m are established according to the conditions of the internal structure of the gripper according to its dimensions, (see figure 5) [5].

$$A_a = \frac{\pi(D_p)^2}{4} = \frac{\pi(0.012m)^2}{4} = 1{,}130 \times 10^{-4}\,[m^2] \tag{2}$$

$$Q = (1{,}130x10^{-4}m^2)(0.0295\frac{m}{s}) = 3{,}336 \times 10^{-6}[\frac{m^3}{s}] \tag{3}$$

2.9. Magnetic permeability

Ability of a material to attract or allow the passage through itself of a magnetic field equation (4), μ_0 is the vacuum magnetic permeability equal to $4\pi\times10^{-7}$ NA^{-2} and μ_r represents the relative magnetic permeability of the 100 NA^{-2} core as in (see figure 5) [7].

2.10. Magnetic hysteresis curve

It shows the relationship between the induced magnetic flux intensity (B) and the magnetization force (H), resulting in a magnetic field of 0.938 Tesla inside the cylinder, equation (5); and a smaller magnetic field in the periphery Bx of 0.0192 Tesla, as in equation (6) [8].

$$\mu = \frac{B}{H} \tag{4}$$

$$B = \frac{\mu_0\mu_r Ni}{L_s} \times N_c \tag{5}$$

$$B_x = \frac{\mu_0\mu_r Ni}{2\pi z} \times N_c \tag{6}$$

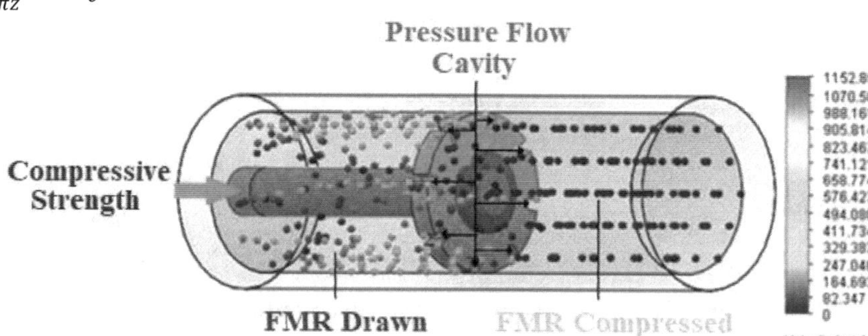

Figure 5. Fibrillar arrangement of the FMR in the shock absorber

2.11. Testing and Simulation

Five cushioning trials were conducted, in each of which values were taken for subsequent data tabulation and evaluation of averages. A simulation of the magnetic hysteresis was also performed in the Comsol Multiphysics software, entering the design parameters of the coil and the density of the fluid equal to 3.15 g cm^{-3} at a temperature of 130 °C according to the data of the fluid manufacturer, to analyze the behavior of the magnetic field of the coil [9].

3. Results

The values obtained in the Tests are reflected in the following tables, where the magnetic flux intensity B and the attenuation force F of the shock absorber have very high values as it rises to a maximum peak current reaching 0.7 amps supplied in steps of 0.1 amps. Table 1 represents the values obtained in the ascent, descent with north polarity at the top and south at the bottom of the coil. Meanwhile in table 2 it represents values at the moment of reversing the polarity, these are affected in the orientation of the magnetic field lines, numerically they change of negative sign in relation to table 1, with the permeability the intensity of magnetic field H is calculated.

Figure 6 shows the FMR magnetic hysteresis curve B-H, which starts from a demagnetized state, where the current starts from zero in the center of the graph. As the magnetic force H increases positively, the flux density B follows the curve from the origin to (point 1) where +B is maximum and saturation is met; while H decreases, the flux follows the curve from (point 1) to (point 2) through Br, which is the residual induction, this (point 2) represents the amount of flux remaining during the falling edge as the magnetizing force decreases; in the third and fourth quadrant the magnetizing force -H where the electromagnet begins to lose its magnetism, as shown by following the curve from (point 2) to (point 3). This point is known as coercive force -Hc where the coil will be completely demagnetized; the curve corresponding to (point 3) and (points 4) is known as -Bmax which represents the saturation peak in the opposite direction to B max; from (point 4) to (point 5) the value -H decreases, this point is known as Flux -Br which represents a magnetization force equal to zero with less magnetic flux; lastly from (point 5) to (point 6) is known as coercive force + Hc reducing the magnetization of the electromagnet to zero and the magnetic force continues to increase until completing the curve from the final point to the initial one [8].

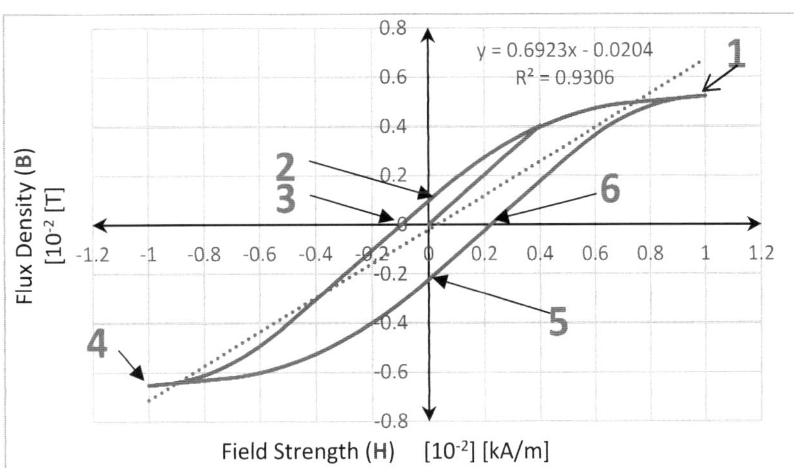

Figure 6. Magnetic Hysteresis Curve

Table 1. Direct Polarization

	I	Test # 1		Test # 2		Test # 3		Test # 4		Test # 5		Average				H
	[A]	B	F	B	F	B	F	B	F	B	F	B		F		kA/m
		Gauss	Kgf	Gauss	Kgf	Gauss	Kgf	Gauss	Kgf	Gauss	Kgf	Gauss	Tesla	Kgf	N	
Increasing	0	-5,14	0,2	-5,21	0,21	-5,09	0,2	-5,22	0,21	-5,12	0,2	-5,156	-5E-04	0,20	2,000556	-3,22E-08
	0,1	19,13	0,28	19,81	0,29	17,59	0,27	25,55	0,32	17,59	0,27	19,934	0,002	0,28	2,804701	1,245E-07
	0,2	37,82	0,36	38,13	0,37	37,81	0,35	40,45	0,41	37,79	0,34	38,4	0,0038	0,36	3,589233	0,0000002
	0,3	52,61	0,43	53,64	0,4	51,98	0,41	55,39	0,53	52,42	0,48	53,208	0,0053	0,45	4,412992	3,325E-07
	0,4	72,51	0,54	73,28	0,54	68,76	0,53	73,72	0,62	69,7	0,53	71,594	0,0072	0,55	5,413270	4,474E-07
	0,5	76,53	0,58	79,42	0,6	75,73	0,57	80,89	0,62	76,49	0,54	77,812	0,0078	0,58	5,707470	4,863E-07
	0,6	91,8	0,65	93,21	0,66	89,98	0,66	95,83	0,66	90,56	0,66	92,276	0,0092	0,65	6,452775	5,767E-07
Max Amplitude	0,7	100,2	0,67	100,8	0,67	98,96	0,68	100,9	0,68	100,91	0,67	100,37	0,01	0,67	6,609682	6,273E-07
Decreasing	0,6	90,23	0,61	91,44	0,62	87,24	0,59	92,34	0,64	89,97	0,6	90,244	0,009	0,61	6,001669	5,640E-07
	0,5	79,86	0,52	80,6	0,56	78,55	0,5	82,65	0,57	78,75	0,51	80,082	0,008	0,53	5,217137	5,005E-07
	0,4	73,11	0,47	73,31	0,47	67,7	0,45	74,73	0,49	72,1	0,46	72,19	0,0072	0,46	4,589512	4,511E-07
	0,3	63,32	0,41	64,18	0,42	59,07	0,37	65,06	0,42	60,53	0,41	62,432	0,0062	0,40	3,981499	3,902E-07
	0,2	50,93	0,34	51	0,36	50,12	0,34	51,88	0,36	50,12	0,32	50,81	0,0051	0,34	3,373487	3,175E-07
	0,1	32,6	0,25	33,45	0,27	29,14	0,23	36,93	0,28	31,56	0,24	32,736	0,0033	0,25	2,490889	2,046E-07
	0	11,82	0,2	12,32	0,21	10,6	0,18	12,55	0,23	10,92	0,19	11,642	0,0012	0,20	1,980943	7,276E-08

Table 2. Inverse Polarization

	I	Test # 1		Test # 2		Test # 3		Test # 4		Test # 5		Average				H
	[A]	B	F	B	F	B	F	B	F	B	F	B		F		kA/m
		Gauss	Kgf	Gauss	Kgf	Gauss	Kgf	Gauss	Kgf	Gauss	Kgf	Gauss	Tesla	Kgf	N	
Increasing	0	10,5	0,2	10,49	0,19	11,42	0,19	9,98	0,18	10,56	0,17	10,59	0,00105	0,18	1,824036	1,207E-05
	0,1	-12,3	0,27	-12,3	0,26	-14,05	0,25	-11,47	0,24	-12,5	0,22	-12,52	-0,00125	0,24	2,432049	-2,85E-05
	0,2	-32,54	0,34	-31,64	0,33	-38,18	0,32	-29,72	0,29	-33,56	0,28	-33,12	-0,00331	0,31	3,059674	-0,000151
	0,3	-49,62	0,38	-49,22	0,37	-51,86	0,36	-48,33	0,36	-50,98	0,3	-50,00	-0,00500	0,35	3,471554	-0,000456
	0,4	-63,29	0,43	-62,41	0,42	-68,89	0,42	-62,21	0,4	-64,78	0,39	-64,31	-0,00643	0,41	4,040339	-0,001173
	0,5	-72,08	0,47	-71,93	0,46	-73,8	0,45	-69,44	0,45	-73,27	0,44	-72,10	-0,00721	0,45	4,452219	-0,002630
	0,6	-79,42	0,53	-77,6	0,52	-82,63	0,51	-73,96	0,5	-75,99	0,48	-77,92	-0,00779	0,50	4,981778	-0,005685
Max Amplitude	0,7	-93,42	0,63	-92,3	0,61	-96,19	0,65	-90,54	0,64	-93,92	0,59	-93,27	-0,00932	0,62	6,119349	-0,006805
Decreasing	0,6	-86,13	0,58	-78,56	0,55	-87,21	0,52	-78,93	0,47	-86,14	0,46	-83,39	-0,00833	0,51	5,060231	-0,003042
	0,5	-73,8	0,45	-72,08	0,44	-77,74	0,43	-71,44	0,45	-74,78	0,42	-73,96	-0,00739	0,43	4,295312	-0,001349
	0,4	-69,3	0,41	-68,56	0,4	-70,92	0,39	-60,65	0,38	-70,31	0,37	-67,94	-0,00679	0,39	3,824593	-0,000619
	0,3	-55,34	0,36	-52,13	0,35	-59,71	0,35	-46,58	0,33	-58,89	0,32	-54,53	-0,00545	0,34	3,353874	-0,000248
	0,2	-44,76	0,31	-43,16	0,28	-46,48	0,27	-42,19	0,27	-45,7	0,26	-44,45	-0,00444	0,27	2,726248	-0,000101
	0,1	-27,49	0,25	-26,36	0,25	-28,95	0,24	-24,6	0,23	-28,11	0,22	-27,10	-0,0027	0,23	2,333982	-3,089E-0
	0	-7,02	0,22	-6,86	0,22	-7,33	0,22	-5,26	0,21	-7,04	0,21	-6,702	-0,00067	0,21	2,118236	-3,820E-0

The analysis of the attenuation force generated in the damper is reflected in the graph corresponding to the fluence stress as shown in figure 7, representing the gradual increase of the numerical value with respect to the magnetic flux intensity, obtaining the maximum attenuation value of the damper of F equal to 6.1193496 N, with a flux density of B equal to 0.01 T. The analysis of the attenuation force generated in the damper is reflected in the graph corresponding to the fluence stress, representing the gradual increase of the numerical value with respect to the magnetic flux intensity, obtaining the maximum attenuation value of the damper of F equal to 6.1193496 N, with a flux density of B equal to 0.01 T.

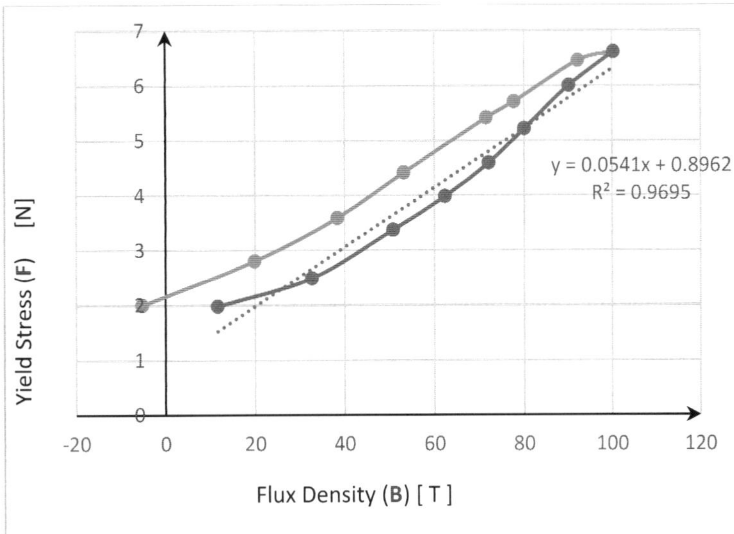

Figure 7. Performance Yield Stress Curve vs. Flux Density

When comparing the curve of the magnetic hysteresis obtained with the tabulated data with respect to the curve obtained by simulation, it is observed that the periodic increase of the force of magnetization H causes that the magnetic flux that circulates through the nucleus is greater than the remaining flux of the exterior during the rising and falling flanks forming this way the hysteresis loop. Finally, we can conclude that we have an FMR hysteresis curve with high permeability, low residual magnetic flux, and low coercive force figure 8 [9].

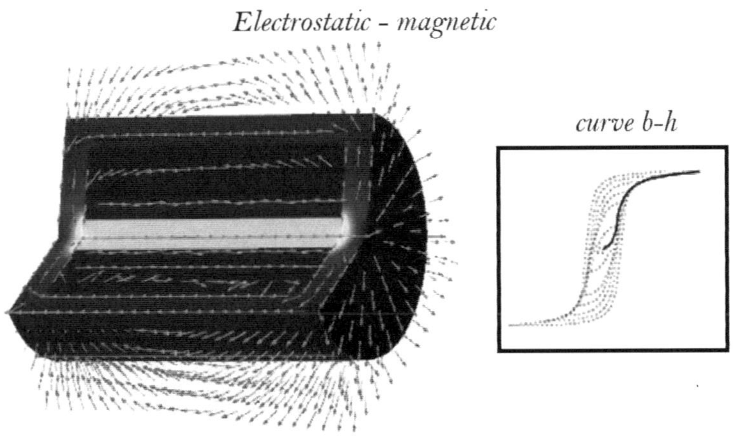

Figure 8. Magnetic field simulation in COMSOL Multiphysics Software

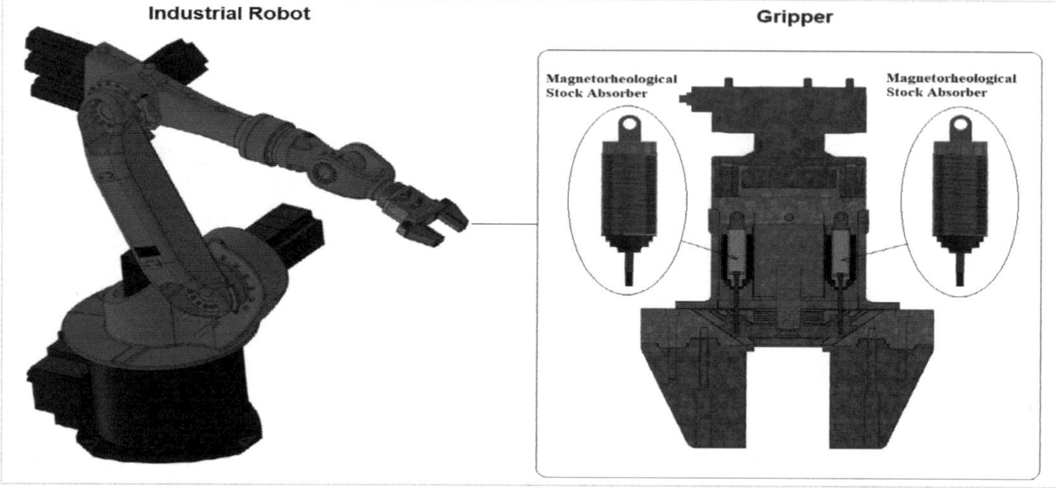

Figure 9. Magnetorheological shock absorber in robotic arm grippers.

4. Conclusion

Research indicates that the magnetorheological fluid produced by LORD MRF-140CG series can be used in the design of industrial shock absorbers, because the maximum attenuation force obtained is 6.6 Newtons, at a current of 0.7 amps applied to the winding, which demonstrates the applicability of the shock absorber in industrial grippers. The innovation of this work is reflected in the design and development of a damping system, based on magnetorheological fluids for grippers or tweezers, which, when installed in industrial robotic arms, will allow safer and more reliable handling of fragile objects such as glass, ceramics and containers of dangerous products, which may otherwise be harmful to the operator. The gripper model with this damping system is not available in the national market, figure 9.

5. References

[1] Kumar, A., & Mangal, S. K. (2013). Modeling testing and evaluation of magneto-rheological shock absorber. *Int J Mech Eng Rob Res*, *2*(1), 71-80.

[2] Diaz, P., Palanikumar, K., & Kumar, P. (2014). Modern Achievements and Developments in Manufacturing and Industry. *Advanced Materials Research, 984-985.*

[3] Boada, M. J. L., Boada, B. L., & López, V. D. (2005). Algoritmo de aprendizaje por refuerzo continuo para el control de un sistema de suspensión semi-activa. *Revista Iberoamericana de Ingeniería Mecánica*, *9*(2), 77.

[4] Cúpich Rodríguez, M., & Elizondo Garza, F. J. (1998). Amortiguadores magnetoreológicos. *Ingenierías*, *1*(2), 50-54.

[5] Walker, S. M. (2007). *El magnetismo.* ediciones Lerner.

[6] Chacón Hernando, V. (2010). *Diseño de una suspensión para un vehículo automóvil basada en amortiguadores magneto-reológicos* (Bachelor's thesis).

[7] Navarro, M. A. G., González, V. A. G., Rivera, M. H., Melo, M. E. R., & Castro, A. T. (2010). Desarrollo de nanopartículas magnéticas en templetes biopoliméricos. *Ingenierías*, *13*(46), 23.

[8] de Blas, A., Bargalló, R., & de la Hoz, J. (2004). Modelado del Ciclo de Histéresis Mediante el Modelo de Preisach. *Información tecnológica*, *15*(3), 31-38.

[9] Mesquida, C. D., & Lässig, J. L. (2012). Introducción a los fluidos magnetoreológicos con aplicaciones en ingeniería.

IOP Conf. Series: Materials Science and Engineering **841** (2020) 012011 doi:10.1088/1757-899X/841/1/012011

Identification of Unbalance in Rotating Machinery Using Vibration Analyse Solution

Phuoc Vinh Dang*, Le Hung Toan Do, Nhu Thanh Vo, Thanh Nghi Ngo, Hoai Nam Le.

Department of Mechanical Engineering, The University of Danang – Danang University of Science and Technology

Email: dpvinh@dut.udn.vn

Abstract. Vibrations in rotating machinery are normally the consequence of mechanical faults such as mass unbalance, manufacturing, mechanical looseness, assembly, installed machines, coupling misalignment, and many other causes. However, unbalance is the most popular reason of machine vibration. An unbalanced rotor always causes more vibration and generates excessive force in the bearing area and reduces the life of the machine. In this paper, experimental studies were performed on a rotor with two mounted disks to identify the unbalance in rotor using order tracking technique based on one key phasor. After calculation of the coefficients of influence α of each experimental test, the matrix of influence coefficient [C] and the global matrix of coefficients [H] can be obtained. Finally, the two suitable masses are identified in order to reduce the unbalance of the rotor systems.

1. Introduction

Rotor unbalance is the most popular reason in rotating machinery. A very small amount of unbalance may cause severe problem in high speed rotating machines. The vibration caused by unbalance may destroy critical parts of the machine, such as bearings, seals, gears and couplings. As a result of mass unbalance, a centrifugal force is generated and must be reacted against by bearing and support structures. The purpose of rotor balancing is to attain satisfactory running when installed on site. It means no more than an acceptable magnitude of vibration is caused by the unbalance remaining in the rotor. In the case of a flexible rotor, it also means that not more than an acceptable magnitude of deflection occurs in the rotor at any speed up to the maximum service speed. Some analytical methodologies have been carried out to unbalance response such as the transfer method. Further, the unbalance part of the rotor rotates at the same speed as the rotor and therefore the force caused by the unbalance is synchronous [1]. Vibration signatures are commonly used as a useful method for investigating progressive machine mechanical malfunctions and form the baseline signature for further comparative monitoring to detect mechanical faults [2]. G.K. Yamamoto [3] proposed a smart experimental setup for vibration measurement and imbalance fault detection in rotating machinery. By adding a correction mass at the location indicated by the system, the vibration of the machine will be effectively reduced. In this paper, experimental studies were performed on a rotor with two mounted disks to identify the unbalance in rotor using order tracking technique based on one key phasor. After calculation of the coefficients of influence α of each experimental test, the matrix of influence coefficient [C] and the global matrix of coefficients [H] can be obtained. Finally, the two suitable masses are identified in order to reduce the unbalance of the rotor systems.

2. Test rig description

Figure 1 depicts an overview of the experimental apparatus, highlighting the main parts of the test rig. Basically, it is encompassed by a flexible rotor-bearing system. The test rig [4] consists of a rigid shaft of 10mm diameter supported by two rolling element bearings at the ends of shaft. The shaft is driven by a 150 W electric motor through a flexible coupling. The motor speed is adjusted by a speed controller with 2 rpm accuracy and can reach a maximum value of 5000 rpm. The disk with a diameter and thickness of 75 mm and 25 mm respectively is installed in the middle of the shaft for the balancing the system. In order to measure the vibration of the shaft, two proximity probes of 90° phase angle are installed at each bearing position in the horizontal and vertical direction. An additional proximity probe is used as a key phasor ([5]-[8]).

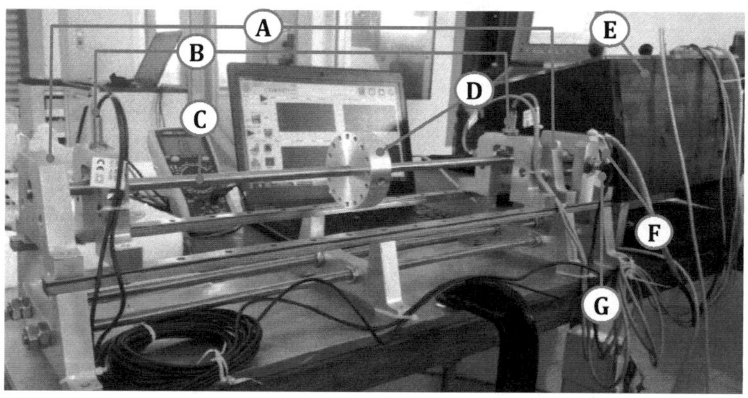

Figure 1. Picture of the test rig: (A) Bearing; (B) Proximity probes; (C) Shaft; (D) Disk; (E) Control box; (F) Electric Motor; (G) Keyphasor [4]

3. Description of experiment procedure

Figure 2 shows the methodology for identification the unbalance of the rotor system. Suppose that two unbalance disks are mounted on the rotating shaft.

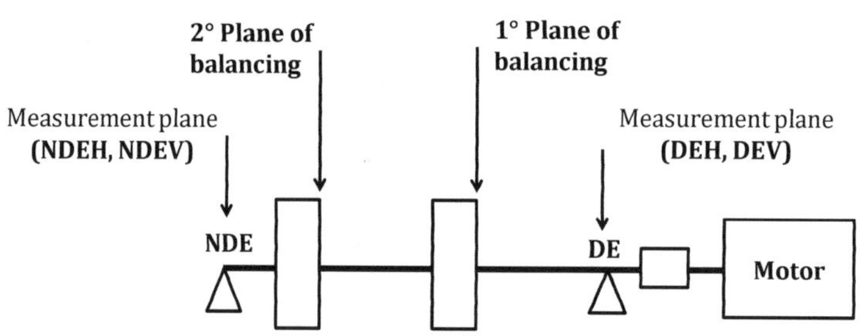

Figure 2. Description of the experimental test.

Firstly, the rotor is run at a speed of 4000 rpm. Next, the run-down test is performed to speed of 500 rpm. This test is known as the reference test. After that, an arbitrary mass is added to the disk #1 (1° balancing plane) and disk #2 (2° balancing plane). The run-down tests are repeated. The following steps are the procedure of the experiment tests:

1. Measure the vibrations of the rotor for one or more unbalanced rotation speeds, on different planes of measurement in the x and y direction (reference tests).

2. Add a known mass arbitrary at the 1° balancing plane.
3. Measure the new vibrations $X^*(\Omega_j)$ at 1° balancing plane.
4. Calculate the coefficients of influence $\alpha(\Omega_j)$ for the *j-th* rotation speed.
5. Repeat steps 2-3-4 for the 2° balancing plane.
6. Assembly the matrix of influence coefficient **[C]**.
7. Obtain a global matrix of coefficients **[H]**.
8. Estimate the 2 masses of imbalance that reproduce the experimental vibration of the rotor.
9. Place masses in opposite position to balance the rotor (add 180° to the phase).

Step. 1: After measuring the rotor vibrations in the time domain, performing order tracking and extracting 1X component $X_z(\Omega_j)$ of the rotor. The velocity vector is $\{\Omega\} = \{\Omega_1 \dots \Omega_i \dots \Omega_N\}^T$ and the 1X component corresponding to each speed is $X(\Omega_j) = \{X_{NDEx}(\Omega_j), X_{NDEy}(\Omega_j), X_{DEx}(\Omega_j), X_{DEy}(\Omega_j)\}^T$

Step. 2: By adding a known mass to arbitrary w_i balancing plane, the force due to the imbalance U_w can be obtained by:

$$U_w = m_w r w e^{i\varphi w} \tag{1}$$

$$F_w = m_w r \Omega^2 e^{i(\varphi w + \Omega t)} \tag{2}$$

where m_w, r_w, φ_w is mass, radial distance from the axis and angular position (phase) of mass

In this test rig, the masses available are 0.1, 0.2, 0.4, 0.8, 1.0, 1.2, 1.6, 2, and 2.2 grams and the positions available on the disks are 0 - 360° in steps of 22.5°.

Step. 3: Measure the new vibrations $\mathbf{X}^*_{zw}(\Omega_j)$ with the added mass

$$X^*_w(\Omega_j) = \{X^*_{NDEx,w}(\Omega_j), X^*_{NDEy,w}(\Omega_j), X^*_{DEx,w}(\Omega_j), X^*_{DEy,w}(\Omega_j)\}^T \tag{3}$$

the difference $\mathbf{X}^*_w(\Omega_j)$ - $\mathbf{X}(\Omega_j)$ vibrations due to the imbalance between the original plus additional mass $\mathbf{X}^*_w(\Omega_j)$ and vibration due to imbalance only original $\mathbf{X}(\Omega_j)$ represent the vibrations from one additional mass.

Step. 4: Calculate the coefficient of influence $a_{zw}(W_i)$ for the *j-th* rotation speed:

$$\alpha_{zw}(\Omega_j) = \frac{X^*_{zw}(\Omega_j) - X_z(\Omega_j)}{U_w} = \frac{X^*_{zw}(\Omega_j) - X_z(\Omega_j)}{m_w r_w e^{i\varphi_w}} \tag{4}$$

The coefficient of influence is the *z-th* vibration 1X exclusively due to a unified force with zero phase acting on the *w-th* balancing plane. For the first plane, we have:

$$\alpha_{NDEx,1}(\Omega_j) \quad \alpha_{NDEy,1}(\Omega_j) \quad \alpha_{DEx,1}(\Omega_j) \quad \alpha_{DEy,1}(\Omega_j) \tag{5}$$

Step. 5: Repeat the steps above to obtain the coefficients of influence for the second balancing plane

$$\alpha_{NDEx,2}(\Omega_j) \quad \alpha_{NDEy,2}(\Omega_j) \quad \alpha_{DEx,2}(\Omega_j) \quad \alpha_{DEy,2}(\Omega_j) \tag{6}$$

Step. 6: Building the global matrix of influence coefficients for the *j-th* rotation speed $[\mathbf{C}(\Omega_j)]$. Under hypothesis of linearity of the system, for the *j-th* rotation speed, we have:

$$\begin{Bmatrix} Y_{NDEx}(\Omega_j) \\ Y_{NDEy}(\Omega_j) \\ Y_{DEx}(\Omega_j) \\ Y_{DEy}(\Omega_j) \end{Bmatrix} = \begin{bmatrix} \alpha_{NDEx,1}(\Omega_j) & \alpha_{NDEx,2}(\Omega_j) \\ \alpha_{NDEy,1}(\Omega_j) & \alpha_{NDEy,2}(\Omega_j) \\ \alpha_{DEx,1}(\Omega_j) & \alpha_{DEx,2}(\Omega_j) \\ \alpha_{DEy,1}(\Omega_j) & \alpha_{DEy,2}(\Omega_j) \end{bmatrix} \begin{Bmatrix} m_1 r_1 e^{i\varphi_1} \\ m_2 r_2 e^{i\varphi_2} \end{Bmatrix} \tag{7}$$

where $Y_z(W_j)$ is the vibration due to the additional mass *w-th* applied measured in the z-axis.

Step. 7: Repeat the previous steps at various speeds to obtain N matrices $[\mathbf{C}(\Omega_j)]$, then be assembled into a single global matrix of coefficients **[H]**

$$\begin{aligned}
&\{\mathbf{y}\} = [H]\{\theta\} \\
&\{\mathbf{y}\} = \{\{\mathbf{y}(\Omega_1)\} \ \dots \ \{\mathbf{y}(\Omega_j)\} \ \dots \ \{\mathbf{y}(\Omega_{ns})\}\}^T \\
&[H] = [[C(\Omega_1)] \ \dots \ [C(\Omega_j)] \ \dots \ [C(\Omega_{ns})]]^T \\
&\{\theta\} = \begin{Bmatrix} m_1 r_1 e^{i\varphi_1} \\ m_2 r_2 e^{i\varphi_2} \end{Bmatrix}
\end{aligned} \tag{8}$$

Step. 8: Matrix **[H]** is the experimental model of the system for different speeds. This matrix allows to simulate the response 1X system for two different masses placed on two distinct planes balancing. The two balancing masses to be placed will have the same value of those of unbalance estimated, the same radial distance and an angular position offset by 180° from these positions. The estimation problem is overdetermined (more equations than unknown parameters):

$$\{\mathbf{y}_{\text{rotor to be balanced}}\} = [\mathrm{H}]\{\theta_{\text{unbalance masses}}\} \tag{9}$$

Assumed an estimate for the masses of unbalance $\{\boldsymbol{\Theta}\}$, vibration simulated obtained with the estimated model **[H]** will be different from the imbalance measured experimentally $\{\mathbf{Y}\}$. The residual value is defined as the difference between quell vibration measured experimentally and the simulated:

$$\{\mathrm{r}\} = \{\mathrm{Y}\} - \{\hat{\mathrm{Y}}\} = \{\mathrm{Y}\} - [\mathrm{H}]\{\hat{\theta}\} \tag{10}$$

The solution of the problem is overdetermined obtained through a least squares approach, minimizing the sum of residues:

$$\{\hat{\theta}\} = \min_{\{\theta\}}\left(\sum_{i=1}^{n} \mathrm{r}_i^2\right) = \min_{\{\theta\}}\left(\{\mathrm{r}\}^T\{\mathrm{r}\}\right) \tag{11}$$

The estimate of the masses of unbalance is obtained through the pseudo inverse of the global matrix **[H]** coefficients of influence:

$$\{\hat{\theta}\} = \left([\mathrm{H}]^{*T}[\mathrm{H}]\right)^{-1}[\mathrm{H}]^{*T}\{\mathrm{Y}\} \tag{12}$$

4. Results

Figure 3 shows the 1X components of four proximity probes (NDEH, NDEV, DEH, DEV) of three tests (reference test and tests with added mass #1 and mass #2). It is clearly that the test with added mass #2 have very large 1X components for all proximity probes.

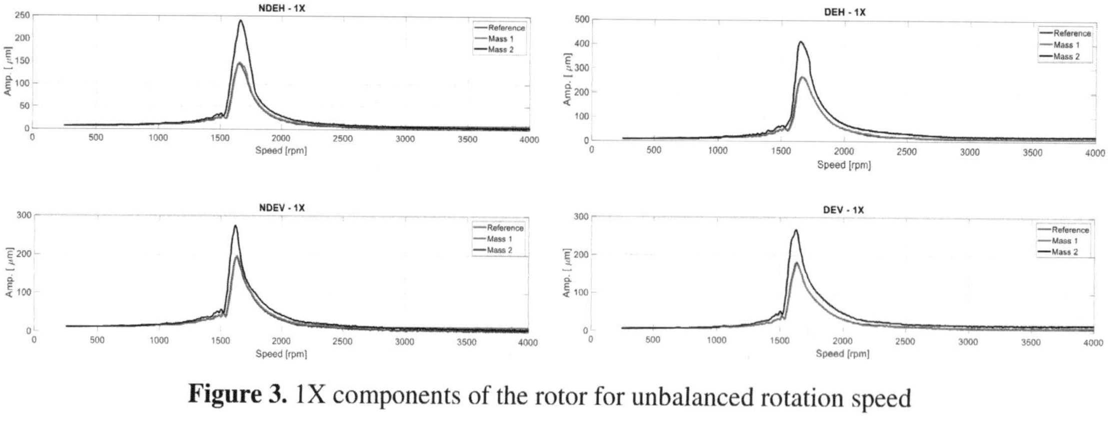

Figure 3. 1X components of the rotor for unbalanced rotation speed

IOP Conf. Series: Materials Science and Engineering **841** (2020) 012011 doi:10.1088/1757-899X/841/1/012011

In order to obtain this 1X components, the acquired signals were applied synchronous averaging after performing order tracking by using a tacho signal. The main advantage of this method is that the noise of 1X signal can be reduced by 70%.

Figure 4 shows the influence coefficients α_1 and α_2 at each proximity probes from speed of 500 rpm to 4000 rpm. Finally, by applying the algorithm presented above step, two suitable masses and their positions can be obtained as: mass #1 = 0.00063kg @ 147° and mass #2 = 0.00088kg @ 288°.

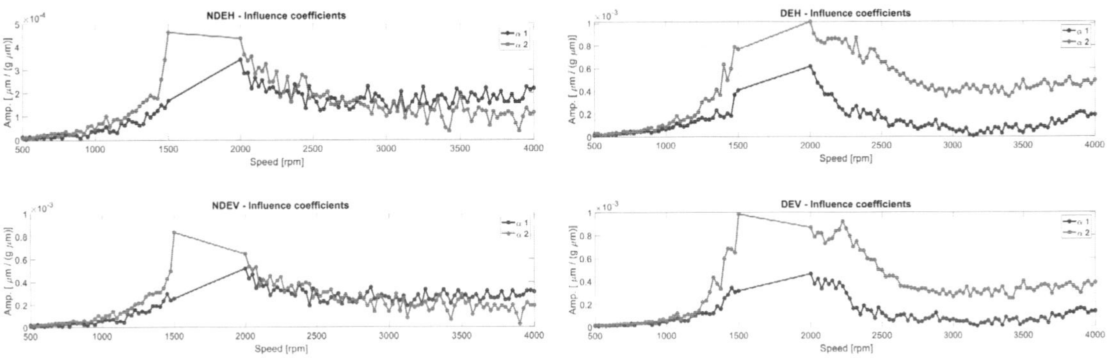

Figure 4. Influence coefficients at four proximity positions

Figure 5 shows the 1X components of four proximity probes before and after balancing as a function of rotor speeds. It is clearly that the 1X components have been significantly reduced for all proximity probes, especially in the range of resonance.

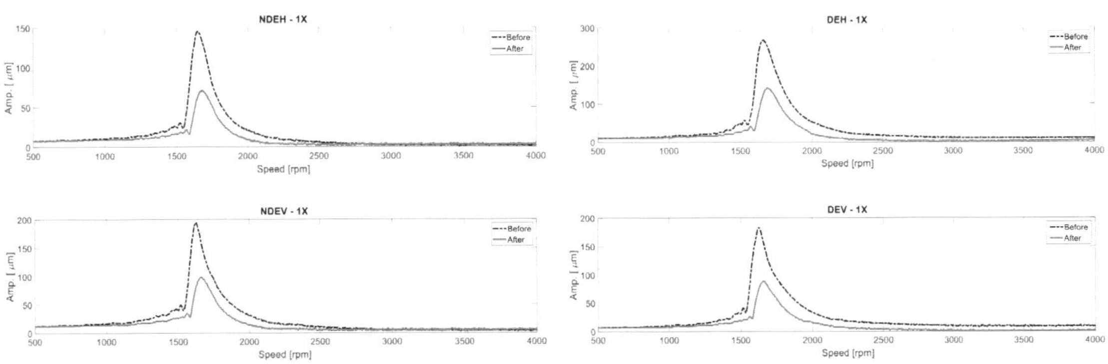

Figure 5. 1X components of the rotor before and after balancing vs. rotation speeds

The proposed experimental setup for vibration measurement and imbalance fault detection in rotating machinery was successfully tested and is ready for application in "real world" systems.

5. Conclusions

An experimental test was performed on a small test rig with two mounted disks to identify the unbalance in rotor using order tracking technique based on one key phasor. The rotor is first run at speed of 4000 rpm and then, the run-down test is performed to 500 rpm. The coefficients of influence $\boldsymbol{\alpha}$ of each experimental test, the matrix of influence coefficient [C] and the global matrix of coefficients [H] can be obtained in order to identify the suitable masses including the value and the position to reduce the unbalance of the rotor systems. This paper only presented the run-down test. The run-up test should be carried out to compare the results between two tests.

References

[1]. Bently D E 2002. Fundamentals of Rotating Machinery Diagnostics. *Bently Pressurized Bearing Press.*

[2]. Ramachandran K P 2004. Vibration signature analysis for machine health monitoring and fault diagnosis. *Caledonian Journal of Engineering* 26-39.

[3]. Yamamoto G, Costa C, and Sinohara J 2016. A smart experimental setup for vibration measurement and imbalance fault detection in rotating machinery. *Case Studies in Mechanical Systems and Signal Processing* **4** 8–18

[4]. Huy P N Q, Vinh D P, Nghi N T and Thanh T P 2018. Design of a small-scale test rig for rotating machinery characterization. *Lecture Notes in Networks and Systems* **63** 229-235.

[5]. Dang P V, Chatterton S, Pennacchi P, and Vania A 2016. Effect of the load direction on non-nominal five-pad tilting-pad journal bearings. *Tribology International* **98** 197-211.

[6]. Chatterton S, Dang P V, Pennacchi P, Luca A D and Flumian F 2017. Experimental evidence of a two-axial groove hydrodynamic journal bearing under severe operation conditions. *Tribology International* **109** 416-427.

[7]. Dang P V, Chatterton S, and Pennacchi P 2019. The effect of the pivot stiffness on the performances of five-pad tilting pad bearings. *Lubricants* **7** 61.

[8]. Chatterton S, Pennacchi P, Dang P.V. and Vania A 2014. A test rig for evaluating tilting-pad journal bearing characteristics. *Mechanisms and Machine Science* **21** 921-930.

Acknowledgement

This work was supported by University of Danang, University of Science and Technology, code number of Project: T2019-02-42.

IOP Conf. Series: Materials Science and Engineering **841** (2020) 012012 doi:10.1088/1757-899X/841/1/012012

Energy-efficient Scheduling for Multistage Manufacturing System with Parallel Machines

Chang-pu MA[1,2], **Bing-hai ZHOU**[1], **Yi SU**[1]

[1]School of Mechanical Engineering, Tongji University, Shanghai 201804, PR China
[2]School of Management, Guilin University of aerospace technology, Guilin, PR China

Email: smcp1108@foxmail.com

Abstract. Increasing energy is prompting academicians to pay attention to energy efficient scheduling. This study establishes a mathematical model for multistage manufacturing system with parallel machines to minimize total energy consumption. A memetic algorithm based on differential evolution(MABDE) is developed to solve the model effectively. Two rules are incorporated into the proposed algorithm to narrow down the search space and improve search efficiency. Computational results demonstrate the effectiveness of the proposed model and algorithm.

1. Introduction

For the manufacturers, introducing power-efficient equipment [1] or designing embodied energy framework can be practical method for decreasing carbon footprints and energy consumption [2]. But compared to the aforementioned method, energy-efficient scheduling (EES) requires less capital investment [3].

In this paper, EES is incorporated into multistage manufacturing system (MMS) with parallel machines aiming at minimizing the total energy consumption. MMS with parallel machines, also known as hybrid flow shop scheduling problem (HFSP), is a generalization of the traditional flow shop scheduling problem (FFSP). Regarding FFSP with EES, Fang et al. [3] proposed a mathematical model for FSP with peak power load and verified the model by applying a case study. Luo et al. [4] presented an ant colony optimization for HFSP considering electricity consumption cost. Lin et al. [5] presented an integrated model for processing parameter optimization and HFSP, and adopted a teaching-learning-based optimization (TLBO) for minimizing make span and carbon footprint. Ding et al. [6] devised a modified multi-objective iterated greedy algorithm for a permutation flow shop scheduling with carbon emission. Lei et al. [7] developed a novel TLBO algorithm for HFSP taking into account the scheduling, machine assignment, and speed selection.

However, the above-mentioned literature did not consider the energy consumption of processing different types of products and that when machines are idle, which are the very issue covered in this study. Moreover, to solve this problem, a MABDE is proposed. MABDE incorporates into the rule of first available equipment (RFAE) and minimal energy consumption equipment (RMECE) to reduce the search space and improve search efficiency.

2. Problem Description

The problem can be depicted as follows. There are N jobs that need to be processed at M stages sequentially from stage 1 to stage M. The mth stage consists of M_m parallel machines, and $M_m \geq 1$ at each stage. Each job n consists of a series operation O_{mn}, m=1, 2,…, M. The actualization of O_{mn} needs one machine out of the machine set M_m at stage M. $L_m=\{l_m\}$ is a random processing sequence of all jobs at the mth stage, where l_m denotes that the job's processing order is l_m at stage m. P_{nmj} is the processing time of O_{mn} on machine j. B_{nmj} is the start time of O_{mn} on machine j at the stage m, while F_{nmj} is the completion time of O_{mn} on the machine j at stage m. EC_{nmj} denotes the unit energy consumption of O_{mn} on machine j at stage m, and EI_{mj} represents the unit energy consumption in unit time when machine j is idle at stage m.

To formulate the model, the following assumptions are considered: The process is not interrupted; Each machine can process only one job at a time, and each job can be processed on only one machine at a time. Each job is independent of each other and can be processed at initial time; The processing time of each job on each machine is known; Transportation times of job between stages and unloading time of job on machine are neglected; The failure of machines is not considered throughout the production process, and the buffer capacity between adjacent stages is unlimited.

3. Mathematical Model

To formulate the problem clearly, two binary variables are introduced. x_{nlm}=1 denotes that O_{mn} is processed in the l_mth order at stage m;0, otherwise. y_{nmj}=1 denotes that O_{mn} is processed on machine j at stage m;0, otherwise. The mathematical model can be formulated as follows.

$$\text{Objective} \quad \min \ EA = ECA + EIA \tag{1}$$

s.t.

$$\sum_{l_m=1}^{N} x_{n,l_m} = 1, \ n = 1,2,\cdots,N \tag{2}$$

$$\sum_{n=1}^{N} x_{n,l_m} = 1, \ l_m = 1,2,\cdots,N \tag{3}$$

$$\sum_{j=1}^{M_m} y_{n,m,j} = 1, \ n = 1,2,\cdots,N; \ m = 1,2,\cdots,M \tag{4}$$

$$F_{nmj} = B_{nmj} + P_{nmj}, n = 1,2,\cdots,N; \ m = 1,2,\cdots,M; \ j = 1,2,\cdots,M_m \tag{5}$$

$$F_{nmj} \leq B_{n(m+1)j'}, n = 1,2,\cdots,N; \ m = 1,2,\cdots,M-1; \ j = 1,2,\cdots,M_m; \ j' = 1,2,\cdots,M_{m+1} \tag{6}$$

$$\sum_{n=1}^{N} x_{nl_m} B_{n1j} \leq \sum_{n=1}^{N} x_{n(l_m+1)} B_{n1j}, l_m = 1,2,\cdots,N-1; \ j = 1,2,\cdots,M_m \tag{7}$$

$$\begin{gathered}\sum_{n=1}^{N} x_{nl_{m_1}} y_{nmj} F_{nmj} \leq \sum_{n=1}^{N} x_{nl_{m_2}} y_{nmj'} B_{nmj'} + L\left(1 - \sum_{n=1}^{N} x_{nl_{m_2}} y_{nmj'}\right), \\ l_{m_1}, l_{m_2} = 1,2,\cdots,N; \ l_{m_1} \leq l_{m_2}; \ m = 1,2,\cdots,M; \ j, j' = 1,2,\cdots,M_m\end{gathered} \tag{8}$$

$$ECA = \sum_{n=1}^{N}\sum_{m=1}^{M}\sum_{j=1}^{M_m} y_{nmj} EC_{nmj}, n = 1,2,\cdots,N; \ m = 1,2,\cdots,M; \ j = 1,2,\cdots,M_m \tag{9}$$

$$EIA = \sum_{m=1}^{M}\sum_{j=1}^{M_m}\left\{\left(F_{mj} - \sum_{n=1}^{N} y_{nmj} P_{nmj}\right) \times EI_{mj}\right\}, n = 1,2,\cdots,N; \ m = 1,2,\cdots,M; \ j = 1,2,\cdots,M_m \tag{10}$$

$$x_{nl_m} = 0,1, n, l_m = 1,2,\ldots,N; \ x_{nmj} = 0,1, n = 1,2,\ldots,N; m = 1,2,\ldots,M; j = 1,2,\ldots,M_m \tag{11}$$

The objective function (1) minimizes the energy consumption. Constraint (2) indicates that the job n is processed once at the m-th stage. Constraint (3) represents that there is only one job at the l_m-th workstation at the m-th stage. Constraint (4) ensures that the job n is processed by only one machine at the m-th stage. Constraint (5) denotes the completion time of job n at the j-th machine at the m-th stage. Constraint (6) involves job-sequencing at the same machine. Constraint (7) indicates that the jobs scheduled in front takes precedence over the later jobs to begin processing at the first stage. Constraint

(8) guarantees that when the jobs are scheduled on a machine in the same stage, the later job must start after the completion of former job, where L is a large number. Constraint (9) indicates the total energy consumption when all the jobs are processed in whole stages. Constraint (10) represents the total energy consumption when the machines are idle in the process of production. Constraint (11) defines the domain of decision variables.

4. MABDE Algorithm

In this paper, differential evolution is applied to the mutation of the memetic algorithm and greedy algorithm is used for local search, thus solving the problem of hybrid flow shop energy scheduling problem.

4.1. Coding and Decoding

Integer coding is used for the problem, and a chromosome represents a solution. For instance, there are five jobs to be processed in the system, a chromosome {2,5,1,3,4} represents that the processing order is $job_2 \rightarrow job_5 \rightarrow job_1 \rightarrow job_3 \rightarrow job_4$ in the first stage. The first come first serve (FCFS) rule is applied in the following stage. Additionally, the rules of combining the available machine with the minimum energy consumption is applied as follows.

- RFAE. When a job selects a machine at a certain stage, the machine that can start the next job earliest is preferred.
- RMECE. When multiple machines satisfy the RFAE, the machine with minimal energy consumption is selected.

4.2. Population Initialization

Let *PS* be the population size, the initial population is randomly generated for two accounts: one is to avoid precocity, the other is to spread the initial solution into the whole solution space to the maximum extent.

4.3. Evaluation Fitness

For the *p*-th chromosome $Chrome_p$, a boundary construction method is applied to convert objective function value *EA* to fitness shown as follows.

$$Fit\left(f\left(p\right)\right)=\begin{cases} I-EA\left(Chrome_p\right) & EA\left(Chrome_p\right)\leq I \\ 0 & otherwise \end{cases} \tag{12}$$

where *I* is a sufficiently positive value.

4.4. Mutation Mechanism

An energy *differential* evolution (EDE) combined with the traditional insertion mutation is adopted to keep the excellent meme fragments as far as possible in the mutation process. The EDE can be defined as follows.

$$\Delta_{EA}\left(k\right)=EA\left(posi_1, posi_2,\cdots, posi_k\right)-EA\left(posi_1, posi_2,\cdots, posi_{k-1}\right), 2\leq k\leq N \tag{13}$$

$$posi_{mu}=\arg\max_{k}\left\{\Delta_{EA}\left(k\right)\right\}, 2\leq k\leq N \tag{14}$$

where $posi_k$ indicates the *k*-th meme of chromosome $Chrome_p$, and $posi_{mu}$ denotes meme location in which the mutation occurs. The $posi_{mu}$ is defined as the bottleneck location of energy consumption.

4.5. Crossover Mechanism

Due to the job sequence-based coding, a generalized order crossover (GOX) is adopted to *ensure* that the chromosomes are in the solution space after crossover. The specific steps are as follows.

Step 1: two paternal chromosomes $Chrome_{p1}$ and $Chrome_{p2}$ are randomly selected from population.

Step 2: a meme fragment *with* length *CL* is selected from $Chrome_{p1}$ and placed it to the first *CL* location of offspring chromosome $Chrome_{p1}$.

Step 3: another meme fragment that is different from that from step 2 is selected from $Chrome_{p2}$, and fill it successively into the location of offspring chromosome $Chrome_{p1}$ from CL+1 to N to generate the offspring chromosome $Chrome_{p1}$.

The generation of another offspring chromosome $Chrome_{p2}$.is the same as $Chrome_{p1}$.

4.6. Selection Mechanism

The roulette-wheel method is adopted to select the chromosome. It uses a probability distribution for selection in which the selection probability of a given schedule is proportional to its fitness.

4.7. Framework of the Proposed MABDE

Based on the previous illustration, the overall framework of the proposed MABDE is provided by Figure 1.

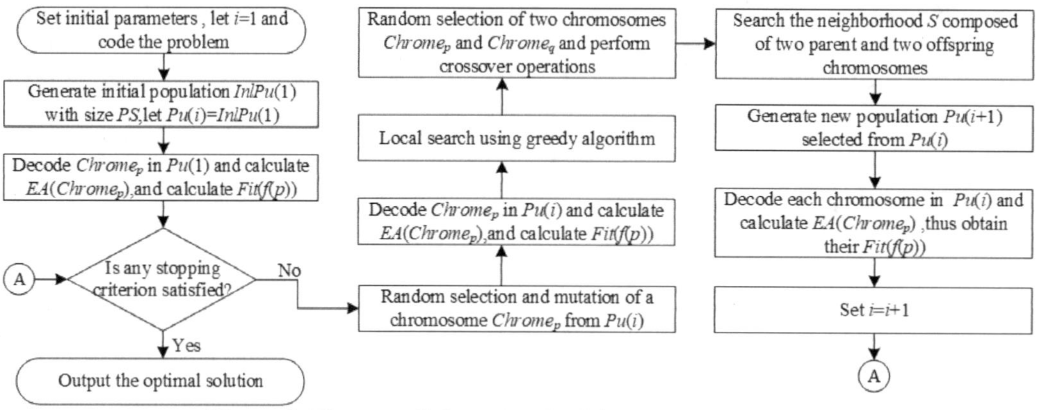

Figure 1. The overall framework of the proposed MABDE

5. Experimental results and analysis

To evaluate the performance of the proposed MABDE algorithm, this section provides an experimental evaluation of its performances and a comparative study with differential evolution (DE) and genetic algorithm (GA), considering significant benchmarking instances of different dimensions. All the algorithms are coded in MATLAB (R2014a) and run on a PC with 2.6 GHz and 4 GB of RAM memory, and Windows 10.

5.1. Experimental setup

Two different scale instances are generated randomly. A small-scale instance is shown in Table 1. For the medium instance, 12 jobs (N=12) need to be processed in four stages (M=4). The alternative machines in each stage is M_1=3, M_2=3, M_3=2, M_4=2. $\forall n \in N, m \in M$. P_{n1j} is generated as $\lfloor rnd^{uni}(45,50) \rceil$,where $rnd^{uni}(45,50)$ is a uniform distribution between 45 and 50, and $\lfloor x \rceil$ aims to convert x into the nearest integer.Additionally, $P_{n2j}=\lfloor rnd^{uni}(30,35) \rceil$, $P_{n3j}=\lfloor rnd^{uni}(30,40) \rceil$, $P_{n4j}=\lfloor rnd^{uni}(25,30) \rceil$; $EC_{n1j}=\lfloor rnd^{uni}(80,100) \rceil$, $EC_{n2j}=\lfloor rnd^{uni}(70,95) \rceil$, $EC_{n3j}=\lfloor rnd^{uni}(70,90) \rceil$, $EC_{n4j}=\lfloor rnd^{uni}(65,90) \rceil$,$EI_{mj}=\lfloor rnd^{uni}(1,4) \rceil$.

For small-scale instance, the population size is set to 20, maximum iteration is 100, mutation probability is 0.001, and crossover probability is 0.7. Meanwhile, the population size is set to 30, maximum iteration is 200, mutation probability is 0.001, and crossover probability is 0.7 for the medium instances.

5.2. Effectiveness Analysis for MABDE

Regarding the stochastic nature of the proposed approach, 10 runs are realized for the resolution of the small-scale instance. Then an evaluating indicator ($BR_{2/1}$) is adopted to measure the experimental results of algorithms. The indicator is calculated as $BR_{2/1}=(AVG_1-AVG_2)/AVG_2\times 100\%$,where $BR_{2/1}$ indicates the superior ratio of algorithm 2 to algorithm 1, and AVG_1 and AVG_2 indicate the average value for ten runs of algorithm 1 and algorithm 2, respectively. Table 2 summarize the computational results. Table 2 shows that the resulting solution of MABDE outperforms DE and GA, where the *BR* is up to 3.86% and 12.62% to DE and GA, respectively.

Table 1. A small-scale instance

Job(n)	Stage 1		Stage 2		Stage 3	
	$P_{n11}(EC_{n11})$	$P_{n12}(EC_{n12})$	$P_{n21}(EC_{n21})$	$P_{n22}(EC_{n22})$	$P_{n31}(EC_{n31})$	$P_{n32}(EC_{n32})$
1	2 (11)	2 (10)	4 (7)	5 (9)	2 (11)	3 (9)
2	4 (14)	5 (9)	3 (9)	4 (15)	5 (13)	4 (11)
3	6 (11)	5 (11)	4 (9)	2 (10)	2 (11)	5 (12)
4	4 (4)	3 (11)	6 (15)	5 (11)	5 (14)	8 (11)
EI_{mj}	1	2	1	2	1	3

Table 2. Computational results of MABDE, DE and GA for the small-scale instance.

Instance	1	2	3	4	5	6	7	8	9	10	*BR*
MABDE	183	180	184	183	182	181	180	180	181	180	0
DE	192	190	189	189	188	190	188	186	187	185	3.86%
GA	207	204	208	204	205	206	201	201	204	203	12.62%

5.3. Convergence Performance Analysis

In order to verify the convergence performance of MABDE, the three algorithms are executed 10 times independently for the medium-scale instance respectively, and the convergence curve is obtained as shown in Figure 2. Obviously, the proposed MABDE can search the satisfactory solution faster than DE and GA, which confirms that MABDE algorithm has outstanding search performance.

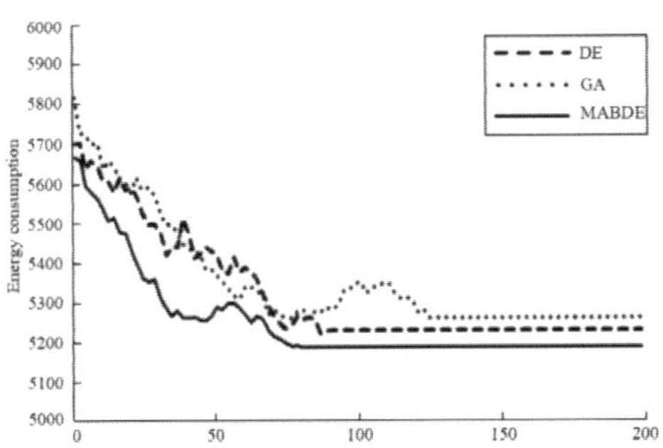

Figure 2. Convergence curve for algorithms.

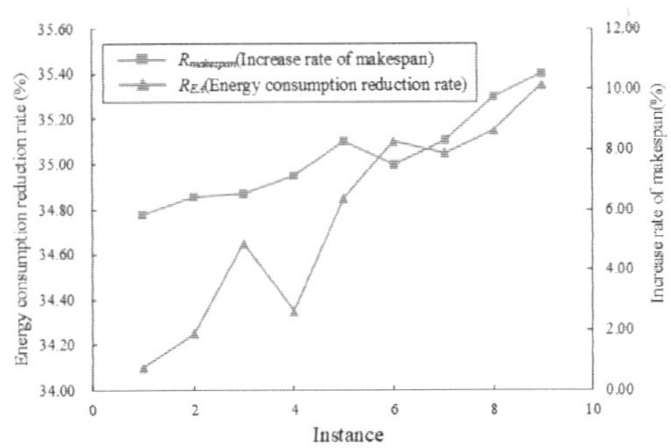

Figure 3. Effect analysis of energy saving strategy

5.4. Energy Saving Analysis

To validate the energy consumption strategy, 9 groups of comprehensive-scale instances are generated randomly. In these instances, 3 groups of jobs(N=6,12,16) are processed in 3 stages (M=2,4,6). Alternative machines in each stage is 3. $\forall n\in N, m\in M$, $P_{nmj}=\lfloor rnd^{uni}(8,15)\rfloor$, $EC_{nmj}=\lfloor rnd^{uni}(30,40)\rceil$, $EI_{mj}=\lfloor rnd^{uni}(3,8)\rceil$. Following parameters are applied in the three algorithms: PS=30, maximum iteration is set to 200, mutation probability is 0.001, and crossover probability is 0.7.

In order to carry on the effective analysis to the proposed energy saving strategy, two types of objective function values are achieved, namely minimum energy consumption and minimum makespan. Moreover, two indicators are adopted as follows.

$$R_{EA} = \frac{EA|G_{Makespan} - EA|G_{EA}}{EA|G_{EA}} \times 100\% \tag{15}$$

$$R_{Makespan} = \frac{Makespan|G_{EA} - Makespan|G_{Makespan}}{Makespan|G_{Makespan}} \times 100\% \tag{16}$$

where R_{EA} is the energy consumption reduction rate, while $R_{Makespan}$ is the increase rate of makespan. $EA|G_{EA}$ indicates the energy consumption with minimizing objective function of EA, while $EA|G_{Makespan}$ denotes the energy consumption with minimizing objective function of makespan. Likewise, the meaning of other symbols in Eq. (16) can be known. Each instance is executed 10 times independently for the three algorithms respectively. Figure 3 shows the results of the energy saving strategy.

In terms of energy consumption, R_{EA} has always between 34% and 35%. And with the increase of the problem scale, the reduction of energy consumption shows an uptrend. From the angle of processing time, $R_{Makespan}$ remains between 5.5% and 6.5% on the small-scale instances. With the increase in the scale of the problem, $R_{Makespan}$ has a slight increase trend. But the maximum value of $R_{Makespan}$ is 10.5% for all instances.

6. Conclusions

This work presents a mathematical model for the multistage manufacturing system scheduling problem with parallel machines for minimizing total energy consumption. To solve this problem, a memetic algorithm based on differential evolution is proposed. The computational results on different scale instances verify the efficiency and effectiveness of the proposed algorithm.

7. References

[1] Mori M, Fujishima M, Inamasu Y and Oda Y 2011 A study on energy efficiency improvement for machine tools *CIRP Ann-Manuf Techn* **60**(1) 145-8.

[2] Kara S, Manmek S, and Herrmann C 2010 Global manufacturing and the embodied energy of products *CIRP Ann-Manuf Techn* **59**(1) 29-32.

[3] Fang K, Uhan N, Zhao F and Sutherland J W 2011 A new approach to scheduling in manufacturing for power consumption and carbon footprint reduction *J. Manuf. Syst.* **30**(4) 234-40.

[4] Luo H, Du B, Huang G Q, Chen H and Li X 2013 Hybrid flow shop scheduling considering machine electricity consumption cost *Int. J. Prod. Econ.* **146**(2) 423–39.

[5] Lin W, Yu D Y, Zhang C, Liu X, Zhang S, and Tian Y, et al 2015 A multi-objective teaching-learning-based optimization algorithm to scheduling in turning processes for minimizing makespan and carbon footprint *J. Cleaner Prod.* **101**, 337-47.

[6] Ding J Y, Song S and Wu C 2015 Carbon-efficient scheduling of flow shops by multi-objective optimization *Eur. J. Oper. Res.* S0377221715004099.

[7] Lei D, Gao L and Zheng Y 2017 A novel teaching-learning-based optimization algorithm for energy-efficient scheduling in hybrid flow shop *IEEE T Eng Manage* 1-11.

Acknowledgement

This study is supported by the Scientific Research Project of Education Department of Guangxi District under the Grant KY2015YB349 and the subject of philosophy and Social Science Planning in Guangxi under the Grant 15FJY008.

IOP Conf. Series: Materials Science and Engineering **841** (2020) 012013 doi:10.1088/1757-899X/841/1/012013

Correlation between Rail Vibration and Sound Radiation Using a Hammer Impact

Bowen Hou[1, 4], Zhuang Shao[2] and Qine Zeng[3]

[1] Department of Road and Railway Engineering, Beijing Jiaotong University, Shangyuancun No.3, Haidian District, Beijing, China
[2] Department of Road and Railway Engineering, Beijing Jiaotong University, Shangyuancun No.3, Haidian District, Beijing, China
[3] Department of Road and Railway Engineering, Beijing Jiaotong University, Shangyuancun No.3, Haidian District, Beijing, China
[4] Corresponding Author: Bowen Hou.

Email: bwhou@bjtu.edu.cn

Abstract. When the track structure is subject to certain excitation, the vibration characteristics and sound radiation characteristics will show certain correlation with each other. However, the study on finding the direct relationship between rail vibration and radiated sound is very limited. In this paper, the correlation between the rail vibration and radiated sound pressure under a hammer impact was investigated with a full-scale track slab. The rail acceleration, rail surface sound pressure and radiated sound pressure nearby at the rail web and rail foot were measured and analyzed in frequency domain. The transfer function, coherence function were obtained to reveal the relevance between each pair of input and output signals. Finally, by choosing rail vibration acceleration and surface sound pressure as input excitation and choosing the surface sound pressure and the sound pressure in near field as output response successively, the transfer function and the transmission loss for each pair of input and output were analyzed and compared to reveal the signal transmission in different frequency ranges.

1. Introduction

As the railway traffic continuously grows in highly populated areas, the environmental vibration and noise induced by moving vehicles has aroused more and more focus [1, 2]. Usually the mitigation track structure [3, 4] and different vibration isolation mats [5, 6] were adopted to reduce the vehicle induced environmental vibration at the vibration source and the transmission path. Rafal Burdzik et al. [7] have conducted experimental test aiming at identifying the vibration characteristics of the railway infrastructure. Antonio D'Andrea et al.[8] have performed a series of vibration measurements at a test section of a railway interconnection in Turin-Milan HS Railway and a quantity of piezoelectric accelerometers were adopted to determine the influence of new hot mix asphalt layer on the vibration transmission. David Thompson et al. [9] have investigated the influence of stiffening the subgrade beneath the railway track on reducing the vibration transmission. Mingjing Fang et al.[10] had presented a model of a Portland cement concrete slab track bed, and a sub- track asphalt roadbed to analyze their responses to ground vibration attenuation under high-speed train loads by considering different thick sub-track asphalt layer replaced on the top of the upper subgrade. However, when the railway mitigation tracks were adopted more and more frequently, certain problem has emerged. It has been discovered that after adopting the ballast mat, although the environmental vibration transmitted to the infrastructure beneath has been reduced, but the radiated sound near the wheel rail contacting zone has been increased significantly, which on the contrary will increase the overall noise radiation [11].

In the past, researchers have conducted theoretical and experimental studies in order to determine the sound radiation by the moving vehicles. Xiaodong Song et al. [12, 13] had studied the sound radiation of the single-span and multi-span bridge by combining the inverse boundary element method and in-site measured data. The model was further verified against the field measurements of a U-shaped girder bridge. Xun Zhang et al.[14, 15] had carried out a field measurement on a simply-supported box girder bridge to measure the accelerations of the slabs and the associated sound pressures induced by running trains and developed a simplified vehicle–track–bridge coupling vibration model to calculate the wheel–rail interaction force in a frequency range of 20–200Hz. From the study above, it can be seen that most of the research focused on the wheel-rail interaction radiated noise, and most regarding the rail to be simulated as a infinite beam and later worked as a line sound source in the simulation However, the railway track structure are consisted of different solid components with complex profiles, thus each small area on the solid surface will give different contributions to the overall sound at certain environment point according to the infinitesimal element method[16–21]. Thus it is more reasonable to study the rail radiated sound through a solid element model. [16–19, 22–24] From the literature above, it can be seen that most researchers are focusing on verifying the numerical model with the near field sound pressure, and little research have been conducted on finding the direct relationship between the rail vibration and sound radiation.

In this paper, the hammer impact load was applied as the system excitation mainly due to the fact that the hammer impact load will provide a wide frequency band of excitation spectra and thus the system response can be regarded as a representation of the track structure characteristics of its own[25, 26]. A laboratory test was carried out for a full-scale track, and the acceleration, the surface sound pressure and the radiated sound pressure near the hammer location have been chosen as evaluation indexes. The responses for each index were analyzed in the frequency domain. Finally, by choosing rail vibration acceleration and surface sound pressure as input excitation and choosing the surface sound pressure and the sound pressure in near field as output response successively, the transfer function and the transmission loss for each pair of input and output were analyzed and compared to reveal the signal transmission in different frequency ranges.

2. Methodology

2.1 General information of the full-scale slab track

The full scale slab track is a typical CRTS II ballastless slab track [27, 28], which is consisted of the rail, the slab, the cement asphalt(CA) mortar and the hydraulically bonded layer(HBL). The sketch of the track slab and full scale model in the lab are shown in Figure 1.

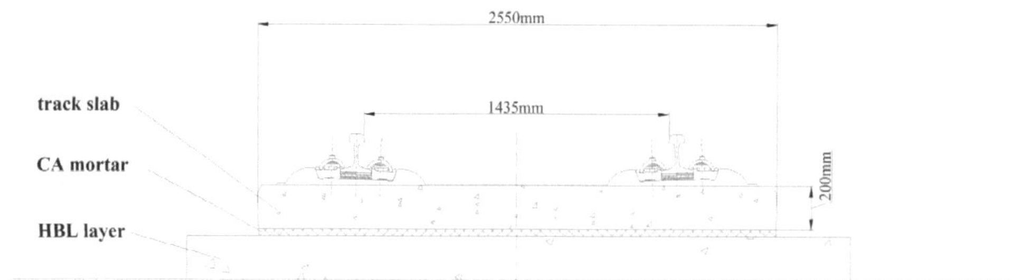

Figure 1. Sketch of the track slab and the full-scale model in the lab.

2.2 Equipment and sensor locations

The data collector is type INV3062C1(L) manufactured by the China Orient Institute of Noise and Vibration. 8 channels of signal can be collected at the same time with the maximum sampling rate of 216kHz and 24 bit precision per channel. Three types of transducers were chosen for the aim of the research, namely are the accelerometers for the rails, surface microphone and 1/2 inch microphones. The technical specification for each type of the transducer is listed in Table. 1. The sampling rate is set to 10.24kHz for each channel.

Table 1. Transducer specification

	Type	Sensitivity	Technical parameter
Accelerometer	ICP	4.72~4.88mV/g	Frequency range: 1 to 15KHz Dynamic range: 0~1000g
1/2 inch microphone	ICP	44.4~49.7mV/Pa	Frequency range: 20～20kHz Dynamic range: 20～146dB
Surface microphone	ICP	11.2 mV/Pa	Frequency range: 5 to 50000Hz Dynamic range: 55~160dB

In this test, all sensors are installed at the mid-span position between two adjacent sleepers along the lateral direction. In together, two accelerometers are installed on the rail web and rail foot. One surface microphone is installed first near the accelerometer at the rail web with 5 cases of hammer test and then installed near the accelerometer at the rail foot mainly because currently there is only one surface microphone ready to work with. Three 1/2 sound microphones are installed near the vibration position, among which two of them are fixed near the accelerometer at the rail web and rail foot, and one in the near field which keeps a distance of 0.35m from the rail web in the lateral position and in the same height as the rail neutral axis. The location of all the sensors are shown in Figure 2. The details of the fixed transducers are shown in Figure 3.

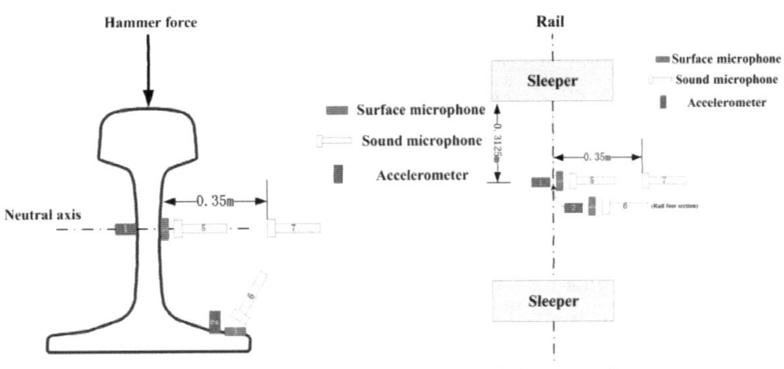

Figure 2. Sketch of the location of the transducer.

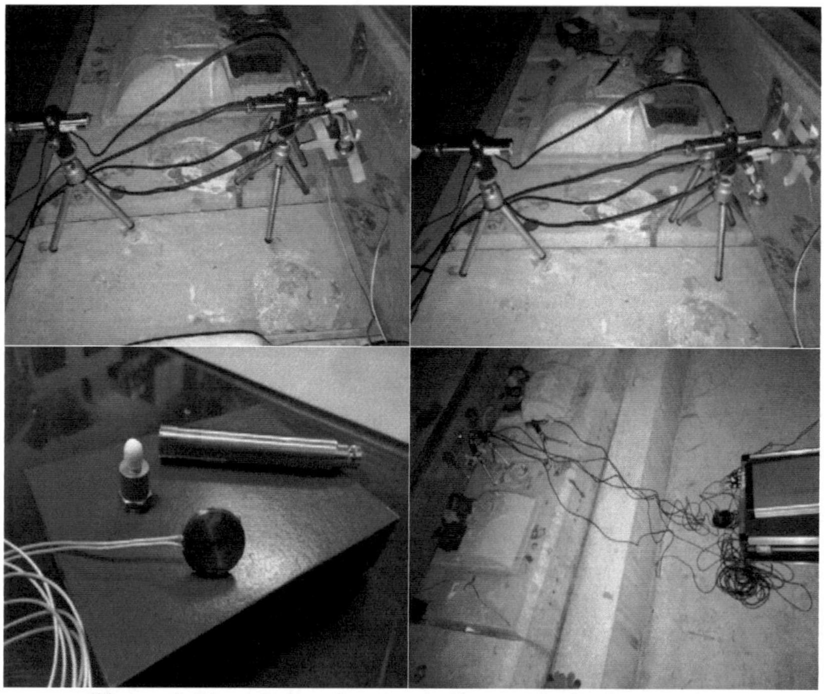

Figure 3. Transducers fixation in the test for both scenarios.

The hammer is directly forced on the rail head along the center line of the rail at the mid-span position between two sleepers. The hammer is able to excite 80kN to the maximum. In order to excite the high frequency response of the rail, the steel hammer head was selected in the test. The sensitivity of the hammer impact force sensor is 0.0639mV/N. All together 5 set of tests were performed, and in each test, the top of the rail was hit three times with a nearly even force with a time lag of 1~2 seconds in order to get a smooth transfer function. By taking the case when the surface microphone is installed at rail foot, the typical hammer forces in five test scenarios and its frequency distribution was shown in Figure 4. It can be seen that with the steel head, the hammer can generate a force with a frequency band from 10~4000Hz with an flat excitation force spectrum, with a nearly 10dB decrease at 4000Hz. Thus it is reasonable to only analyze the results in the following chapters focusing only on the frequency range from 10Hz~4000Hz. However, one thing should be mentioned that this doesn't account for that there are not frequency peaks over 4000Hz. On the contrary, although the excitation force is small comparing with the force below 4000Hz, there are also obvious peaks over 4000Hz, which means the impact load can easily excite the vibration mode with resonance frequency over 4000Hz.

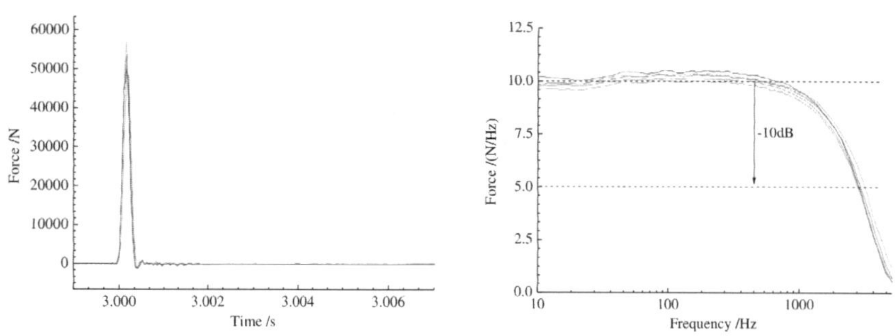

Figure 4. Hammer force in time and frequency domain.

In order to measure the real vibro-acoustics data, it is important to know the background noise before carrying out formal test. According to ISO 11204-2010 and ISO 1680-2013, when the background noise level is at least 10dB lower than the acoustic event, it can be regarded that the background noise can be neglected when the microphone is located at the same position for two measurements[29]. Thus before the formal measurement starts, a test measurement was carried out to obtain the sound level in the presence and absence of the hammer impact to see the difference. Figure 5 shows the comparison between the impact noise and the background noise and it can be seen for all the sound sensors the difference between the impact noise and background noise lies in the range from 34.13 to 62.46dB(A). Thus it can be concluded that the background noise can be neglected in the whole test.

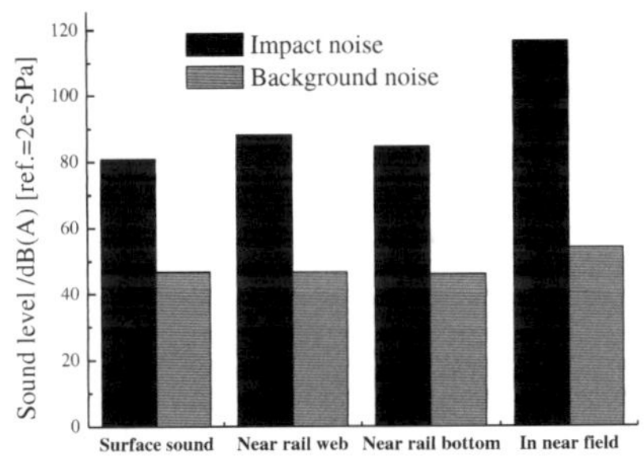

Figure 5. Comparison of sound level between impact noise and background noise.

3. Results and Discussion

The data collector is type INV3062C1(L) manufactured by the China Orient Institute of Noise and Vibration. 8 channels of signal can be collected at the same time with the maximum sampling rate of 216kHz and 24 bit precision per channel. Three types of transducers were chosen for the aim of the research, namely are the accelerometers for the rails, surface microphone and 1/2 inch microphones. The technical specification for each type of the transducer is listed in Table. 1. The sampling rate is set to 10.24kHz for each channel.

The measured accelerations, surface sound pressure and sound pressure in the near field is first analyzed in the frequency domain through Fast Fourier Transformation(FFT) and 1/3 octave analysis according to ISO266-1997. Then the transfer function and coherence function between each pair of signals are obtained and analyzed further. The transfer function is mainly contributed to characterize the system energy transfer capability[30–32] and is defined by equation 1:

$$H(\omega)=\frac{Y(\omega)}{X(\omega)}=\frac{Y(\omega)\cdot X^*(\omega)}{X(\omega)\cdot X^*(\omega)} \tag{1}$$

where $X(\omega)$ and $Y(\omega)$ are the FFT of the input and output signal respectively, $X^*(\omega)$ is the conjugation of $X(\omega)$.

The coherence function[33–35] is usually adopted to evaluate the extent of correlation between each pair of input and output signal. The coherence function is obtained by using equation 2:

$$\gamma_{xy}^2(\omega)=\frac{\left|G_{xy}(\omega)\right|^2}{G_{xx}(\omega)\cdot G_{yy}(\omega)} \tag{2}$$

where $G_{xy}(\omega)$ is the cross power spectral density function between the input and output signal, $G_{xx}(\omega)$ and $G_{yy}(\omega)$ are the self power spectral density function respectively. When the values of coherence function are equal to 1, it means that the relationship between the input and the output are totally correlated to each other without any noise contained. Because the noise in the entire system cannot be avoid, the value of coherence function is always smaller than 1. Usually a coherence function with a value larger than 0.8 means the output signal has strong correlation with the input signal.

In each set of test, the measured data get good convergence in time domain and frequency domain, thus in the following chapter all the curves are combined into one curve using the linear average results at each discrete frequency to give a better illustration.

3.1 Rail acceleration and radiated sound in frequency domain

Figure 6 shows the rail acceleration in both frequency domain and 1/3 octave band. From the results it can be seen that the rail foot vertical acceleration and the rail web lateral acceleration have a different energy distribution in the frequency domain. For the vertical acceleration at rail foot, the main frequency components are at around 487Hz, 895.3Hz, 1535Hz, 2113~2356Hz, 2606Hz, and 2841Hz. Over 3000Hz, the energy get damped and peaks are comparatively smaller. For the lateral acceleration at rail web, the main frequency components are at around 1535Hz, 2574Hz and 3867Hz. According to [20], when the vibration propagation waves in the rail is higher than 3000 Hz (for example 5000, 50,000, and 80,000 Hz in the literature), the rail profile's local deformation becomes the major vibration mode, thus it can be regarded that the hammer impact have excited the local deformation of the rail web and thus there are certain frequency components over 3000Hz. However, for rail foot, the local deformation vibration mode is not excited by the hammer impact.

From the 1/3 octave analysis, it can be seen that at low frequency range from 10~125Hz, the rail web shows a predominant vibration amplitude larger than that at the rail foot, and the same trend can be observed at around 2000~3150Hz. In other frequency band, both curves show nearly the same frequency component distribution. Another thing can be discovered is that although the excitation is only applied on the rail head in the vertical direction, the lateral vibration of the rail web can also be excited.

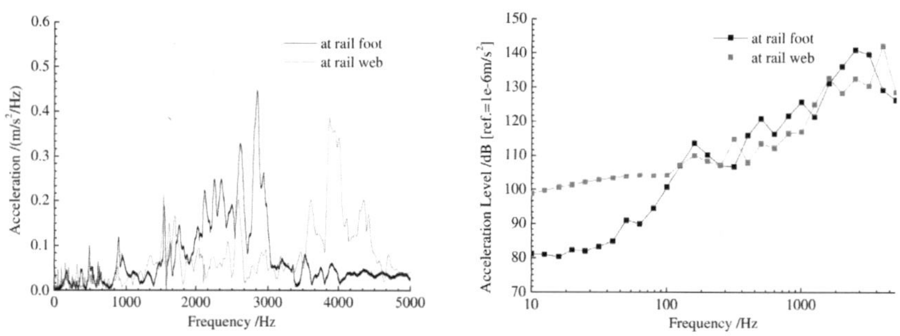

Figure 6. Rail acceleration in frequency domain and 1/3 octave band.

Figure 7 shows the surface sound pressure in both frequency domain and 1/3 octave band. From the results it can be seen that the surface sound pressure at rail foot shows a predominant frequency distribution at 388Hz, 488Hz, 608Hz, 892Hz and 1556Hz. The energy in the range from 1556Hz to 2000Hz is comparatively smaller and has no obvious distribution above 3000Hz. The surface sound pressure at rail web shows a predominant frequency distribution at 1535Hz, 2045Hz and 2578Hz, and the amplitude is smaller than the predominant frequency component of that at the rail foot. According to [36, 37], the track has several resonant frequencies which correspond to the entire structural vibration mode below 3000Hz, such as the pinned-pinned frequency, so it can be concluded that the predominant frequency components of the surface sound pressure generated by the rail foot vibration mainly distributes in the frequency range from 388~1500Hz. On the other hand, the predominant frequency components of the surface sound pressure generated by the rail web vibration mainly distributes in the frequency range from 1535~2578Hz.

From the 1/3 octave analysis, it can be seen that at low frequency range from 10~1250Hz, the surface sound pressure generated by the rail foot vibration shows a predominant sound level with

4.9~20.7dB larger than that at the rail web, and at the frequency range over 1600Hz the surface sound pressure at rail web is larger than that at rail foot.

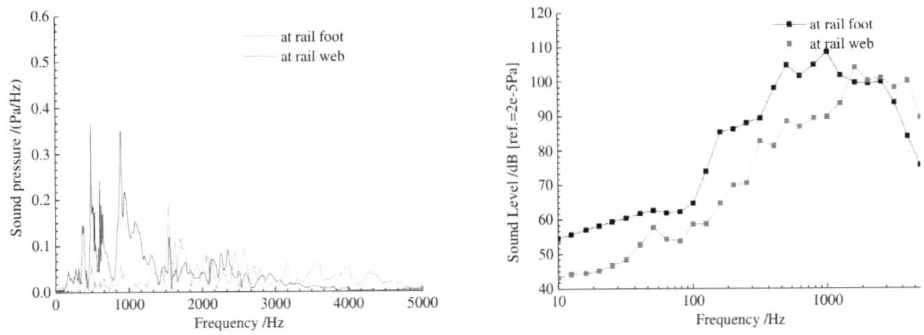

Figure 7. Surface sound pressure in frequency domain and 1/3 octave band.

Figure 8 shows the sound pressure near the accelerometer in both frequency domain and 1/3 octave band. It can be seen the sound pressure near the rail web shows a larger amplitude in almost all frequency range. The predominant frequency range of the sound pressure at rail web is 491Hz, 893Hz,1623Hz, 2115Hz and above 3000Hz. For the sound pressure near rail foot, the predominant frequency range only distributes at around 488Hz and 891Hz. From the 1/3 octave analysis, it can be seen that the sound level near the rail web is 3~14dB larger at each central frequency than that near the rail foot. It can also be observed that the frequency domain curves for both measuring points shows a similar distribution. This means the influence of the surface sound pressures are becoming less predominant after the vibration wave propagated and transmitted into the air volume. In other words, the sound pressure near the solid surface can be regarded as an overall sum up of all the contributions from the infinitesimal sound source at the solid surface and get damped with nearly the same decay rate at each frequency band. In this way the sound pressure at both measuring points show a similar distribution in the frequency domain.

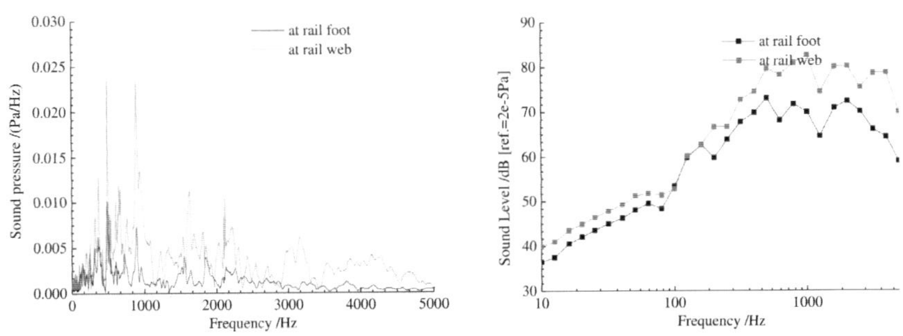

Figure 8. Sound pressure near accelerometer in frequency domain and 1/3 octave band.

Figure 9 shows the sound pressure at near field in both frequency domain and 1/3 octave band which is 35cm away from the rail web surface. From the results it can be discovered that there are several peaks in the frequency domain curve, which namely are at around 365Hz, 487Hz, 1539Hz, 2089Hz and 2288Hz. From the 1/3 octave analysis, it shows several peaks at the central frequency of 160Hz, 400Hz and 1600Hz. The maximum sound level over all central frequencies is 71.5dB.

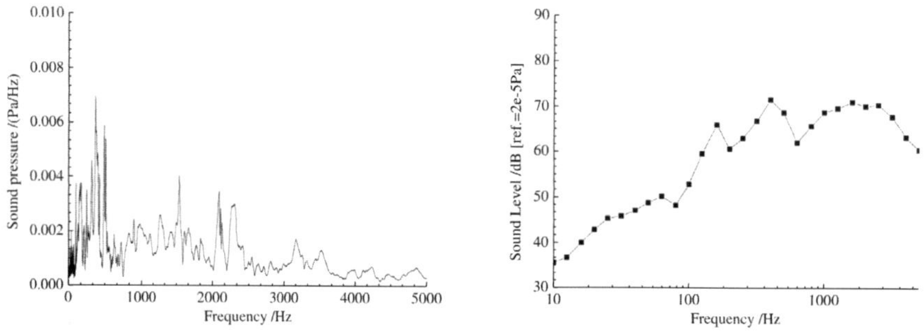

Figure 9. Sound pressure at near field in frequency domain and 1/3 octave band.

3.2 Transfer function and coherence function analysis

Figure 10 shows the transfer function and coherence function between the rail acceleration and surface sound pressure in both frequency domain. From the results it can be seen that for the energy transmission between the rail acceleration and surface sound pressure, in the frequency range from 20~1447Hz the transfer function at rail foot shows a larger amplitude than that at rail web. In the frequency above, the rail web begins to show a better energy transmission than the rail foot. This means with the same acceleration level, the rail vertical vibration can induce the surface sound in low and mid-frequency more easily, and the rail web can induce the surface sound in higher frequency range more easily.

From the coherence function, it can be discovered below 77Hz, the rail acceleration and surface sound generation has no direct relevance with each other. It can also be seen that there are some decreasing points at around 50Hz, 100Hz, 300Hz and 440Hz, which means that there exists little relevance between each pair of input and output signal at around these frequencies.

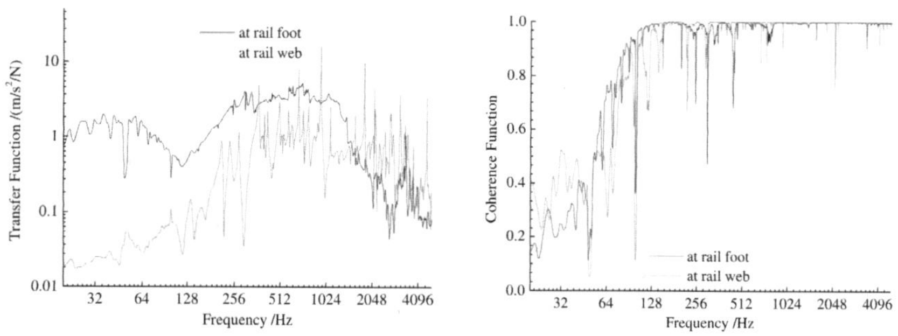

Figure 10. Transfer function and coherence function between surface sound pressure and rail acceleration.

Figure 11 shows the transfer function and coherence function between the surface sound pressure and sound pressure near the accelerometer in both frequency domain. It has clearly shown from the transfer function results that the rail web has a better energy transmission for converting the surface sound to radiated sound into the air nearby, and thus caused a higher amplitude in nearly all the frequency range in the transfer function curve.

From the coherence function, it can be discovered below 101Hz, the rail surface sound and the sound pressure caused in the air nearby has weak relevance with each other, especially for the rail foot. It can also be seen that there are some decreasing points at around 117Hz, 221Hz, 259Hz, 290Hz and 457Hz, which means that there exists little relevance between each pair of input and output signal at around these frequencies.

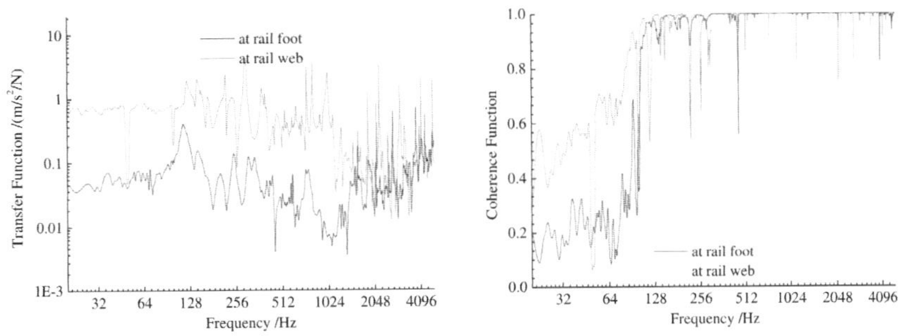

Figure 11. Transfer function and coherence function between sound pressure near accelerometer and at surface.

Figure 12 shows the transfer function and coherence function between the surface sound pressure and sound pressure at near field in both frequency domain. Because the output signal in this analysis is the same for both inputs, the transfer function here can also be regarded as an evaluation of the contribution from each measuring location. It can be seen from the transfer function results that the rail web generally have a larger amplitude in most frequency range, which represents that the rail web basically contributes more to the sound pressure than the rail foot to the point in the near field which has a certain distance from the rail surface.

From the coherence function, it can be discovered below 98Hz, the rail surface sound and the sound pressure in the nearby field has weak relevance with each other, especially for the rail foot. It can also be seen that there are some decreasing points at around 118Hz, 218Hz and 457Hz, which means that there exists little relevance between each pair of input and output signal at around these frequencies.

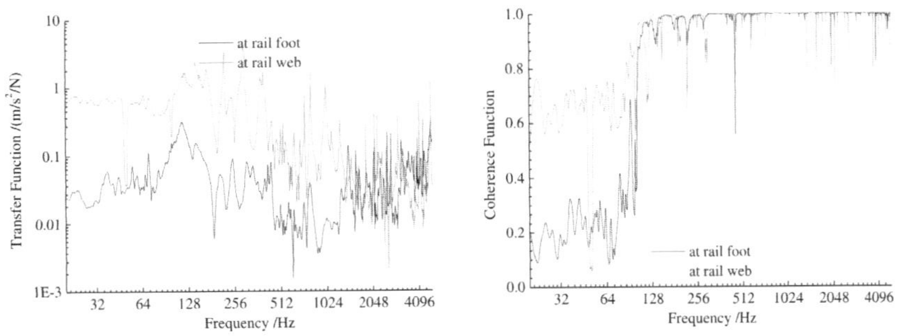

Figure 12. Transfer function and coherence function between sound pressure at near field and at surface.

3.3 Vibration-sound transmission loss analysis

The wave propagation and attenuation from the input excitation to the output response in each transmission path is identified by the transmission loss at each 1/3 octave band central frequency which is defined using equation 3:

$$Transmission\ Loss = 20\times\log(P_{incident} / P_{transmitted}) = 20\times\log(P_{incident}) - 20\times\log(P_{transmitted}) \quad (3)$$

where the $P_{incident}$ refers to the input excitation which namely are the rail vibration acceleration and the surface sound pressure successively; the $P_{transmitted}$ refers to the output response which namely are the surface sound pressure and the sound pressure in near field successively. The transmission loss for each pair of input and output were analyzed and compared to reveal the signal transmission in different frequency ranges.

Figure 13 shows the 1/3 octave band for each measuring point both at the rail foot and rail web. Here, in order to see more clearly the vibration transmission loss, the vibration accelerations have been converted with the same reference which equals 0.00002m/s2 as the sound pressure. From the comparison, it can be found that for the rail foot scenario, the rail acceleration and surface sound pressure are nearly the same below 160Hz. Then the rail surface sound pressure shows a magnitude 8dB~11dB larger than rail acceleration, and above 1600Hz, the surface sound pressure starts to decrease. For both the sound pressure measured at near the accelerometer and in the near field, the 1/3 octave band curve are nearly the same, with a slight difference less than 8dB. For the rail web scenario, the rail acceleration only shows a similar distribution in the frequency range from 400~2000Hz, and in other frequency range the rail web acceleration is larger than the surface sound pressure. In the frequency range from 10~400Hz, the surface sound pressure is nearly the same as the sound pressure in near field measuring points, but the surface sound pressure shows larger amplitudes than the field measuring points, which indicates that the high frequency noise are getting damped through transmission in the air.

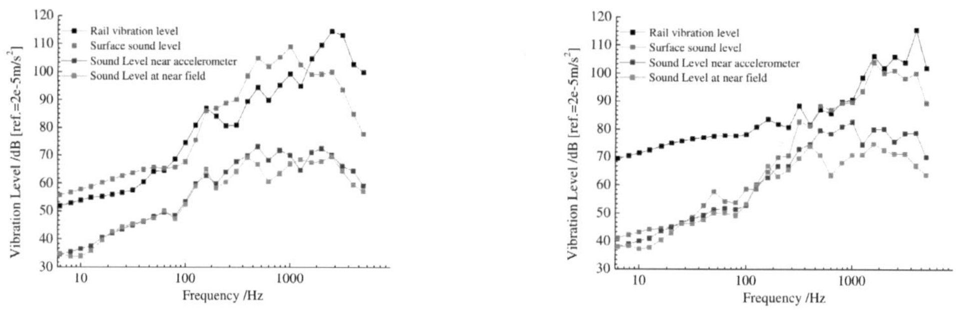

(a) at rail foot (b) at rail web

Figure 13. 1/3 octave band for each measuring point.

Figure 14 shows the transmission loss between each pair of input-output signals at the rail web and at rail foot respectively. By comparing that with the transmission loss between each pair of input and output signals, it can be discovered for the rail foot, the vibration energy loss occurs during the procedure when the surface sound pressure is propagating to the air, and the transmission loss is nearly 14~38dB at different central frequencies. As for the rail web, the vibration energy loss occurs during the procedure when the rail web acceleration induced the surface sound pressure, and the transmission loss is from 5.6~29.8dB at different central frequencies below 400Hz.

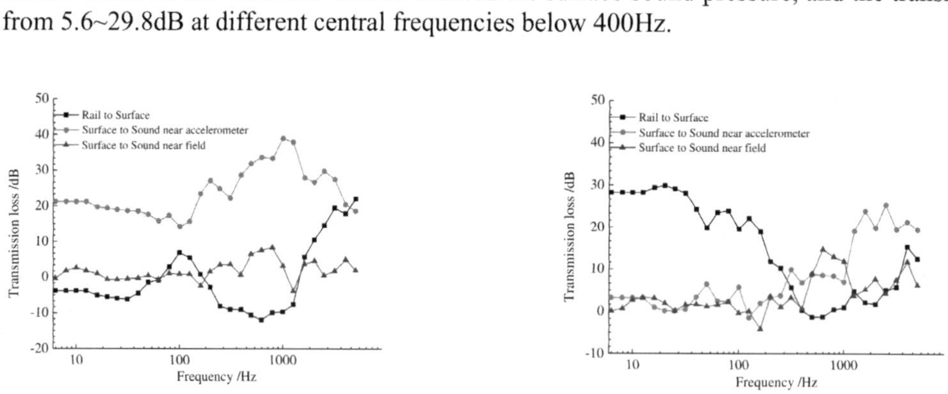

(a) at rail foot (b) at rail web

Figure 14. Transmission loss between each pair of signals.

4. Conclusions and Future Developments

In this paper, a hammer test was conducted for a full-scale track to reveal vibraion propagation and transmission from the rail to the air. The acceleration, surface sound pressure and sound pressure nearby have been measured and analyzed in the frequency domain. By choosing rail vibration acceleration and surface sound pressure as input excitation and choosing the surface sound pressure and the sound pressure in near field as output response successively, the transfer function and the transmission loss for each pair of signals were analyzed and compared to reveal the signal transmission in different frequency ranges The following conclusions can be drawn from this study.

When the excitation is applied on the rail head in the vertical direction, the lateral vibration of the rail web can also be excited and have a predominant energy distribution above 3000Hz. The rail foot only shows a predominant energy distribution below 3000Hz. The predominant frequency of the surface sound pressure generated by the rail foot vibration mainly distributes in 500~1500Hz. The predominant frequency of the surface sound pressure generated by the rail web vibration mainly distributes in 1500~2500Hz.

The influence of the surface sound pressures becomes less predominant after the vibration wave propagated and transmitted into the air. With the same acceleration level at the rail foot and rail web, the rail vertical vibration can induce the surface sound in low and mid-frequency below 1400Hz more easily, and the rail web can induce the surface sound in frequency higher than 1400Hz more easily. The rail web has a better transmission efficiency for converting the surface sound to radiated sound into the air nearby than the rail foot, and contributes more to the sound pressure than the rail foot to the field point.

For the rail foot, the vibration transmission loss occurs during the procedure when the surface sound pressure is propagating to the air, for the rail web, the vibration transmission loss occurs during the procedure when the rail web acceleration induced the surface sound pressure.

5. References

[1] Vogiatzis, K., & Vanhonacker, P. (2016). Noise reduction in urban LRT networks by combining track based solutions. Science of The Total Environment, 568, 1344–1354.

[2] Connolly, D. P., Marecki, G. P., Kouroussis, G., Thalassinakis, I., & Woodward, P. K. (2016). The growth of railway ground vibration problems — A review. Science of The Total Environment, 568, 1276–1282.

[3] Dudkin, E. P., Andreeva, L. A., & Sultanov, N. N. (2017). Methods of Noise and Vibration Protection on Urban Rail Transport. Procedia Engineering, 189, 829–835.

[4] Cui, R., Gao, L., Cai, X., & Hou, B. (2014). Vibration and noise reduction properties of different damped rails in high-speed railway. Noise Control Engineering Journal, 62(4), 176–185.

[5] Kraśkiewicz, C., Lipko, C., Płudowska, M., Oleksiewicz, W., & Zbiciak, A. (2016). Static and Dynamic Characteristics of Resilient Mats for Vibration Isolation of Railway Tracks. Procedia Engineering, 153, 317–324.

[6] Diego, S., Casado, J. A., Carrascal, I., Ferreño, D., Cardona, J., & Arcos, R. (2017). Numerical and experimental characterization of the mechanical behavior of a new recycled elastomer for vibration isolation in railway applications. Construction and Building Materials, 134, 18–31.

[7] Burdzik, R., & Nowak, B. (2017). Identification of the Vibration Environment of Railway Infrastructure. Procedia Engineering, 187, 556–561.

[8] D'Andrea, A., Loprencipe, G., & Xhixha, E. (2012). Vibration Induced by Rail Traffic: Evaluation of Attenuation Properties in a Bituminous Sub-ballast Layer. Procedia - Social and Behavioral Sciences, 53, 245–255.

[9] Thompson, D. J., Jiang, J., Toward, M. G. R., Hussein, M. F. M., Dijckmans, A., Coulier, P. Lombaert, G. (2015). Mitigation of railway-induced vibration by using subgrade stiffening. Soil Dynamics and Earthquake Engineering, 79, 89–103.

[10] Fang, M., & Cerdas, S. F. (2015). Theoretical analysis on ground vibration attenuation using sub-track asphalt layer in high-speed rails. Journal of Modern Transportation, 23(3), 214–219.

[11] Yi, Q., Wang, P., Zhao, C., Tang, J., & Xu, H. (2017). Spatial Distribution Characteristics and Reduction Measures of Environmental Noise in Elevated Railway Region. Tiedao Xuebao/Journal of the China Railway Society (Vol. 39).

[12] Song, X. D., Li, Q., & Wu, D. J. (2017). Study on structure-borne low-frequency noise from rail transit bridges using inverse boundary element method. Procedia Engineering, 199, 1380–1385.

[13] Song, X. D., Wu, D. J., Li, Q., & Botteldooren, D. (2016). Structure-borne low-frequency noise from multi-span bridges: A prediction method and spatial distribution. Journal of Sound and Vibration, 367, 114–128.

[14] Zhang, X., Li, X., Zhang, J., Song, L., & Li, Y. (2016). A hybrid model for the prediction of low–frequency noise emanating from a concrete box-girder railway bridge. Proceedings of the Institution of Mechanical Engineers, Part F: Journal of Rail and Rapid Transit, 230(4), 1242–1256.

[15] Zhang, X., Li, X., Hao, H., Wang, D., & Li, Y. (2016). A case study of interior low-frequency noise from box-shaped bridge girders induced by running trains: Its mechanism, prediction and countermeasures. Journal of Sound and Vibration, 367, 129–144.

[16] Vlahopoulos, N., Kalish, Y., & Raveendra, S. T. (1999). Development of an Equivalent Force Method and an Application in Simulating the Radiated Noise from an Operating Diesel Engine. Shock and Vibration, 6(3), 113–123.

[17] HAMDI, M. A., & VILLE, J. M. (1982). Development of a sound radiation model for a finite-length duct of arbitrary shape. AIAA Journal, 20(12), 1687–1692.

[18] Sheng, X., Zhong, T., & Li, Y. (2017). Vibration and sound radiation of slab high-speed railway tracks subject to a moving harmonic load. Journal of Sound and Vibration, 395, 160–186.

[19] Zhang, X., Thompson, D. J., & Squicciarini, G. (2016). Sound radiation from railway sleepers. Journal of Sound and Vibration, 369, 178–194.

[20] Ryue, J., Thompson, D. J., White, P. R., & Thompson, D. R. (2009). Decay rates of propagating waves in railway tracks at high frequencies. Journal of Sound and Vibration, 320(4–5), 955–976.

[21] Ryue, J., Thompson, D. J., White, P. R., & Thompson, D. R. (2008). Wave Propagation in Railway Tracks at High Frequencies. In Noise and Vibration Mitigation for Rail Transportation Systems (pp. 440–446). Berlin, Heidelberg: Springer Berlin Heidelberg.

[22] Yang, X., & Shi, G. (2014). The Effect of Slab Track on Wheel/Rail Rolling Noise in High Speed Railway. Intelligent Automation & Soft Computing, 20(4), 575–585.

[23] Peng, S.-H., & Haase, W. (Werner). (2008). Advances in hybrid RANS-LES modelling : papers contributed to the 2007 Symposium on Hybrid RANS-LES Methods, Corfu, Greece, 17-18 June 2007. Springer.

[24] Lyu, Y., Bergseth, E., Olofsson, U., Lindgren, A., & Höjer, M. (2015). On the relationships among wheel–rail surface topography, interface noise and tribological transitions. Wear, 338–339, 36–46.

[25] Asgari, A., Sedighi, M., & Riahi, M. (2015). Investigation of Punching Parameters Effect on Mechanical Properties of Al-1100-O in Incremental Sheet Metal Hammering Process. Strength of Materials, 47(6), 882–889.

[26] Mizui, M., Yamamoto, I., Kimura, S., & Maeda, M. (2017). Research on Hammering Test System by Unmanned Aerial Vehicles for Infrastructure Surveillance (pp. 25–32). Springer, Cham.

[27] Zhang, N., Zhou, S., Xia, H., & Sun, L. (2014). Evaluation of vehicle-track-bridge interacted system for the continuous CRTS-II non-ballast track slab. Science China Technological Sciences, 57(10), 1895–1901.

IOP Conf. Series: Materials Science and Engineering **841** (2020) 012013 doi:10.1088/1757-899X/841/1/012013

[28] Yan, B., Liu, S., Pu, H., Dai, G., & Cai, X. (2017). Elastic-plastic seismic response of CRTS II slab ballastless track system on high-speed railway bridges. Science China Technological Sciences, 60(6), 865–871.

[29] Kotus, J., Lopatka, K., & Czyzewski, A. (2014). Detection and localization of selected acoustic events in acoustic field for smart surveillance applications. Multimedia Tools and Applications, 68(1), 5–21.

[30] Park, J., Ahn, S., Kim, J., Koh, H.-I., & Park, J. (2017). Direct determination of dynamic properties of railway tracks for flexural vibrations. European Journal of Mechanics - A/Solids, 61, 14–21.

[31] Auersch, L., & Said, S. (2017). Dynamic testing of track damage before and after construction, after damage and after repair.

[32] Kouroussis, G., Vogiatzis, K. E., & Connolly, D. P. (2017). Assessment of railway ground vibration in urban area using in-situ transfer mobilities and simulated vehicle-track interaction. International Journal of Rail Transportation, 1–18.

[33] Tirelli, D., & Vadillo, I. (2013). A Fast Automated Impact Hammer Test Method for Modal Parameter Extraction: Implementation on a Composite Bridge Beam. In Nondestructive Testing of Materials and Structures (pp. 999–1006). Dordrecht: Springer Netherlands.

[34] Baqersad, J., Poozesh, P., Niezrecki, C., & Avitabile, P. (2014). Comparison of Modal Parameters Extracted Using MIMO, SIMO, and Impact Hammer Tests on a Three-Bladed Wind Turbine (pp. 185–197). Springer, Cham.

[35] Noguchi, Y., Maruyama, T., Ueno, K., Komai, M., Takeda, N., & Kakudate, S. (2016). Impact hammer test of ITER blanket remote handling system. Fusion Engineering and Design, 109–111, 1291–1295.

[36] Knothe, K., Strzyzakowski, Z., & Willner, K. (1994). Rail Vibrations in the High Frequency Range. Journal of Sound and Vibration, 169(1), 111–123.

[37] Blanco, B., Alonso, A., Kari, L., Gil-Negrete, N., & Giménez, J. G. (2017). Distributed support modelling for vertical track dynamic analysis. Vehicle System Dynamics, 1–24. [1] Sze S M 1969 *Physics of Semiconductor Devices* (New York: Wiley–Interscience)

Acknowledgments
The authors acknowledge the Ministry of Science and Technology of the People's Republic of China [grant number 2017YFB1201104]; the Fundamental Research Funds for the Central University[grant number 2018JBM042] ; the National Natural Science Foundation of China [grant numbers 51708021] and China Railway Corporation [grant number 2016G009-B].

AUTHOR INDEX

Institute of Physics
Dirac House, Temple Back
Bristol BS1 6BE UK

ISSN: 1757-8981
ISBN 978-1-7138-2683-5

CURRAN ASSOCIATES INC.
proceedings.com